SILVER AND ENTREPRENEURSHIP IN SEVENTEENTH CENTURY POTOSÍ

Anonymous portrait of Antonio López de Quiroga

Silver and Entrepreneurship in Seventeenth-Century Potosí

The Life and Times
of Antonio López de Quiroga

Peter Bakewell

University of New Mexico Press
Albuquerque

LIBRARY OF CONGRESS
Library of Congress Cataloging-in-Publication Data

Bakewell, P. J. (Peter John), 1943–
    Silver and entrepreneurship in seventeenth-century Potosí: the
life and times of Anthony López de Quiroga / Peter Bakewell.—1st
ed.
        p. cm.
    Bibliography: p.
    Includes index.
    ISBN 0-8263-1097-4.   ISBN 0-8263-1098-2 (pbk.)
    1. Silver industry—Bolivia—Potosí (Dept.)—History—17th
century. 2. López de Quiroga, Antonio, ca. 1629–1699.
3. Capitalists and financiers—Bolivia—Potosí (Dept.)—Biography.
4. Potosí (Bolivia : Dept.)—History. I. Title.
HD9536.B63P623   1988
338.7 '6223423 '09–dc19
[B]
                                                            88-14227
                                                                CIP

FOR SUSAN

CONTENTS

Illustrations

PREFACE

MY INTEREST IN ANTONIO LÓPEZ DE QUIROGA BEGAN IN POTOSÍ IN the early 1970s, as an offshoot of more general research that I was doing there in the Casa Nacional de Moneda on the early history of that greatest of Spanish American silver mining centers. López's name lives on in present-day Potosí as an epitome of the town's days of ancient wealth and splendor. Sustaining that almost folk-memory of him is a crudely painted but forceful portrait that hangs in the galleries of the Casa de Moneda (see Frontispiece).* No painter's name appears on this picture of López, but the canvas bears three legends, apparently each by a different hand. Each of them is in some way wrong. But the mistakes themselves reflect the strength of the impression that Antonio López made on Potosí.

The least of the errors is in the title at the lower left: *El Maestre de Campo Don Antonio López de Quiroga, Año de* 1660. López was not, in fact, to become a *maestre de campo* (a military rank roughly equivalent to colonel) for another decade after 1660; and his name was never graced with the honorific *don*.

At the upper left, surrounded by triumphal wreath, appear the words, *Fundador de la Iglesia y Convento de San Francisco de Potosí* ("Founder of the Church and Monastery of Saint Francis in Potosí"). This is either great ignorance or conscious exaggeration on the part of whoever painted the message, for the Franciscan monastery of Potosí traces its beginnings to the middle of

*I am grateful to Lincoln Draper for photographing this portrait.

the sixteenth century, a hundred years or so before López came on the scene.

Finally, at the upper right of the portrait, in a third style of lettering, is the statement that López contributed 21 million pesos in *quintos*, or silver royalties, to the Spanish treasury during his career: *Dio Veinte y un millones de Quintos, como consta en la Real Caja de esta Villa de Potosí.* Though López was the leading silver producer of Potosí for most of the part of his life that he spent there, and also probably the preeminent silver miner and refiner of the Spanish American Empire of his day, that figure is still a wild exaggeration. The best estimate that can be made of López's silver output (see the Conclusion below) is that his mines and refineries yielded, at most, just under 15,000,000 ounces during his career. The *quinto* payable on this amount would have been only some 3,400,000 pesos.

The texts on the portrait give us, then, an Antonio López de Quiroga exaggerated in social standing, munificence, and wealth. But these exaggerations can be read as evidence of the expansion of López's figure in Potosí's memory after his death. Different hands added their comments to the canvas as the reality of his life faded from view and the myth began to grow.

As I hope this book shows, however, there was really no need to exaggerate. López was remarkable enough in the ways that the texts on the portrait bring out. And he was equally remarkable in ways that they do not suggest: as an investor, a technical innovator, a backer of explorations, a land owner, a trader, a politician, and perhaps above all as an integrator of all these activities. He must rank, indeed, as one of the most diverse and capable businessmen to appear in the whole span of the Spanish Empire in America.

Once the portrait had caught my attention, then, I began, as a pleasant diversion from copying numbers from early Potosí treasury accounts and other such tedious research chores, to collect whatever fragments appeared on López and his family and associates. Eventually this interest developed, in the archive of the Casa de Moneda in Potosí, into a systematic inspection of as many notarial volumes for the second half of the seventeenth century as I could find. This produced a good haul of contracts, wills, and the odd dowry or two, all of it adding up to a useful outline of López's career. A little later, in the Bolivian National Archive, in Sucre, I found a splendid source of information on López in the large set of lawsuits about colonial mining preserved there. In both archives, and subsequently in the Archive of the Indies in Seville as well, the official correspondence of colonial administrators also proved informative on López, especially for the last three decades of his life, when he was rich and successful enough to attract official attention.

The end product of these archival labors seemed, and still seems, of course, far from enough. I am all too conscious of knowing only bits and pieces of López's life. I am ignorant, for example, of when he was born, and of when he moved from Spain to South America. A general problem that confronts biographers of Spanish colonial figures, that of the absence of personal papers, is certainly a hindrance in this case, too. The nearest thing to personal writing from López that has appeared is a very small number of letters of his to business subordinates that survive buried deep in lawsuits, and two or three representations of his own services that he sent to the crown, in the hope of getting rewards, at various stages in his career. These pieces of self-acclamation seem particularly revealing of him, and I have quoted from them liberally in my text. But journals or letters home to relatives in Spain would have been far more valuable; and of those there are, apparently, none. In Galicia, López's native ground, no sign has yet appeared of a family archive, or even of scattered documentary references to his family, despite my own search in the archives of Lugo, the nearest city to his birthplace, and that of helpful friends and archivists in the central Galician repositories at La Coruña.

It was nevertheless in Galicia that my pursuit of Antonio López de Quiroga came in a sense to its end, in a reverse of the great journey from Spain to America that he had undertaken in his own life. One autumn day in November 1986—a squally afternoon when in good Galician fashion it was either raining or just about to do so—I had the good fortune to be taken in search of López's birthplace by a generous friend, Dr. Juan Gelabert, of the Department of Modern History at the University of Santiago de Compostela. We drove at some speed from Santiago to Lugo, and from there more slowly over the minor roads to Triacastela, the small town in which, according to the record of López's burial still surviving in Potosí, he had been born. We had heard that the parish priest of Triacastela knew of some document that might refer to the Quirogas. The priest we discovered to be up to his elbows in wine making, and unable to take us to see what he had found. This, however, he described as a *tabla*, a wooden board, on which was old lettering among which he had made out the name Quiroga. The board was in the vestry of the chapel belonging to the hamlet of Cancelo, a few miles outside Triacastela.

So on we went to Cancelo, down ever muddier lanes and tracks, and eventually came to the chapel, a squat, grey stone place with a slate roof. And there indeed in the vestry was the *tabla*, a framed piece of thin wood, about three by five feet. It was filled with lettering in capitals, painted in black on a white ground, and mostly in good condition, except toward the bottom where moisture had damaged the wood. We took the panel outside, into better light under the porch, and there, interrupted by the occasional showers that the wind swept by, set about reading and copying the text.

The first line made it clear enough that we were in luck: "En el nombre de IHS dico io Alvaro de Rivera i Quiroga . . ." ("In the name of Jesus, I, Alvaro de Rivera y Quiroga, say . . ."). These are all names in the family background of Antonio López de Quiroga. Reading on, we found that the panel was the record of a endowment , or *vínculo*, being made by Alvaro de Rivera to "his" chapel of Cancelo. Income from land that he owned nearby was settled in perpetuity on the chapel, for its upkeep, for the maintenance of a priest, and for candles and masses to be offered for the well-being of Rivera's soul after his death. No one besides Alvaro de Rivera should be buried in the chapel, except future patrons of it, and then only on condition that they bore the name Rivera. As his immediate heir, and patron of the chapel, Rivera appointed Mendo de Rivera y Velón (we had borrowed the key to the building from the Velón family, who still occupy the "Casa de

Velón" just up the lane.) Rivera ended his bequest with a recitation of family relatives, among them, at three degrees removed,
don Gaspar de Quiroga, Archbishop of Toledo, grandson (he noted) of Constancia García de Quiroga y Losada and Pedro Vásquez
de Somoza. Antonio López later liked to refer to don Gaspar as
one of his distinguished forebears; and Losada and Somoza are
names intimately linked with his career in Potosí. (See Chapters
3 and 4 below.) Also mentioned is the name Reimóndez, the family name of Antonio López's mother.

This Alvaro de Rivera y Quiroga could have been the father of
Antonio López de Quiroga. Antonio López states his father's name
as Alvaro de Quiroga y Rivera (see Chapter 1, below), and the rever-
sal of the surnames might be of no particular significance. The
date at the end of the endowment is not clear; but the best reading
is 1586, especially given the reference to don Gaspar de Quiroga,
who occupied the Archbishopric of Toledo from 1577 to 1594. If
indeed Alvaro was the father, then the endowment recorded on
the panel at Cancelo must have been made when he was a young
man, for Antonio López was probably born three decades or more
later. (Chapter 1, below.) The eternal wellbeing of one's soul is
not perhaps the commonest concern of youth, but creating an
endowment for that purpose is certainly something that a young
man might do in the face of illness or adversity. Also suggestive
of Rivera's youth is that he makes no mention of wife or children in his bequest, and leaves as heir not a direct descendant,
but a relative.

Here, then, is possibly a relic of Antonio López's father as a
young, as yet unmarried, man. That, for present purposes, would
be the most satisfying interpretation of the panel at Cancelo. But
even if it is not the right interpretation, and the maker of the
bequest was some other, older relative, nothing much is lost. What
the panel shows beyond doubt is that a family carrying all the
surnames later used by, and associated with, Antonio López de
Quiroga lived at Cancelo, in the parish of Triacastela, in the
decades immediately before he was born. The family was notable enough on the local scale, at least, to be the patron of the
hamlet's chapel, and rich enough in income from the land to cre-

ate an endowment for it. A few hundred yards from that little church there stands a farmhouse known still as the "Casa de Quiroga." Whether it, or some earlier building on the site, was the birthplace of Antonio López, cannot be said. But the likelihood must be that he came into the world somewhere very near this remote spot among the Galician hills.

I am most grateful to Juan Gelabert for enabling me to follow Antonio López to earth in Galicia, and especially for his kindness in returning later to Cancelo to retake photographs that had not come out well the first time. To him I also owe my introduction to *tarta de Mondoñedo*, the queen of almond confectionery. I am also indebted to his colleague in the Department of Modern History at Santiago, Dr. Pegerto Saavedra, for having driven me around much of southeastern Galicia one luminous July day of 1986. Other friends in the same Department I thank for allowing me to try their patience with talk of Antonio López and questions about the Galicia of his time.

Once more I thank the Director of the Archivo General de Indias, doña Rosario Parra Cala, and her staff, for their help and kindness while I was searching for traces of López in that vast store of manuscripts. I must also recall with gratitude the help of many colleagues and friends at the Escuela de Estudios Hispanoamericanos in Seville. There I was able to offer a *mesa redonda* on López, and received suggestions that have found their way into the present text. Among *sevillano* friends, I should like to single out with special gratitude Cristina García Bernal and Julián Ruiz Rivera, whose welcome always makes Seville seem like home away from home.

The data on López that I gathered in Potosí were collected in the Casa Nacional de Moneda at a time when the museum as a whole was directed by don Armando Alba, and its archive by don Mario Chacón Torres. Both, sad to say, have subsequently died, and I can express only belated thanks to them for their help with this and other research. To hear don Armando's resounding tones echoing in the morning through the patios of the Casa de Moneda, and especially the explosive "Bah!" that punctuated his speech, was to know that the day's work had indeed begun. Another Potosí friend, don Jack Aitken Soux, is fortunately very much still with us, and I should like to record again my gratitude to him for many kindnesses, and much good information and advice about Potosí and its affairs, past and present. One of the pleasantest memories of my two years in Bolivia is of the tranquility and beauty of his house and farm at Cayara, in one of the valleys below Potosí.

Still presiding over the Bolivian National Archive in Sucre, to the great good fortune of historians, is Dr. Gunnar Mendoza L., whose detailed cataloguing of that rich collection makes work on the history of colonial Charcas so much quicker and easier. For that lifetime's labor, and for much specific help, freely offered, with Antonio López and other matters, I thank him profoundly.

Most of my research on Antonio López was done while I held a junior research fellowship from Trinity College, Cambridge. For the expression of confidence implicit in the awarding of a fellowship, and for the material support provided under its terms, I remain beholden to the College. Generous financial backing for my research on Potosí also came from the Social Science Research Council in London. A later visit to Seville, in 1983, for the purpose of completing my survey of documents dealing with López in the Archivo General de Indias, was funded by a Mellon Foundation Travel Fellowship granted by the Latin American Institute of the University of New Mexico.

This book is the main product of an academic year that I spent at the Institute for Advanced Study, in Princeton, New Jersey, as a member of the School of Historical Studies. No environment, physical or intellectual, more conducive to writing could be imagined, and I owe a permanent debt of gratitude to the Director of the Institute, and the Faculty of the School of Historical Studies, for inviting me to Princeton for that 1985–86 year. It was a particular pleasure to be able to work, once more, in close contact with Professor J. H. Elliott, and a great boon to have access to his large private library of works on Spanish and Spanish-American history. In the end, I hope, I would have produced some account of Antonio López de Quiroga. But without the opportunity for unhurried reflection that those months at the Institute provided,

I doubt that a work of book length would have been forthcoming. And a book is no less than Antonio López de Quiroga deserves.

Finally, my deepest thanks must go to my wife, Susan Benforado, for her cheerful understanding and toleration when author's moodiness overcame me, for her sharp-eyed but always helpful observations on the text as it slowly emerged, and, above all, for her support and encouragement in so many aspects of life. This book is hers.

PROLOGUE

SCENES

IT WAS ANTONIO LÓPEZ DE QUIROGA'S DOING THAT THE SLAVE found himself on the scaffold in the plaza of Potosí. To the crowd that had gathered to see the execution that January day in 1657, the punishment seemed perhaps out of proportion to the offence that the man had committed. But, some of them might have reflected, a sensible slave, under whatever provocation, should not have accused a well-connected and ambitious master of conniving at, or perhaps encouraging, such a grave crime against the king as debasement of coinage in the mint. Especially should he not have done so when only a few years before a great scandal about the adulteration of silver coin with copper had led to executions, imprisonments, and the ruin, in business and honor, of many rich and powerful people connected with that same mint. This slave, however, by name Juan Luis Osorio, now revealed himself, in final and dramatic words, to have been far from sensible not only in this episode of his life, but in many others before it. Standing at the top of the ladder to the gallows, he asked Antonio López to forgive him for his accusation, which had been made in anger. He asked forgiveness of the authorities of Potosí for resisting arrest in the mint with a dagger. He had been frightened by the prospect of torture by López, who had been seen entering the building carrying a trestle of the sort used by torturers. He had, he exclaimed, done nothing in Potosí deserving punishment. In Cuzco, though, he had killed seven men in self-defense; and he

I

had smashed with a stone the head of a student who was teaching him to play the organ, and who had said something objectionable to him; and in La Plata he had killed a black woman who was going to poison him. And so, unburdened of these dreadful secrets, at about noon on that summer day, Juan Luis Osorio came to the end of his thirty years. The executioner pushed him off the ladder, clambered onto his shoulders, and thrust down on him many times until it seemed that he was quite dead.[1] It is not recorded whether Antonio López witnessed this execution, whether he felt any remorse about it, or perhaps even satisfaction that he had unknowingly been responsible for the just punishment of a multiple murderer. To any employee, servant, or slave of his who watched, however, the lessons must have been plain. One did not play with Antonio López. Disservice to him was not merely unwise, but likely to bring fierce retribution down on the offender's head.

Some fifteen years pass. Antonio López de Quiroga has prospered in trade, finance, and mining. The imperial government has recently acknowledged his success by bestowing on him the honorary military rank of *maestre de campo*. In the principal church of Potosí, across the plaza from the mint, he has now congregated with his two young daughters, doña María and doña Lorenza, and many other relatives and retainers, to celebrate the good progress of a new enterprise. With López's encouragement and financial support, his nephew, don Benito de Rivera y Quiroga, has completed the first stage of an exploration in the lowlands to the east. The aim of nephew and uncle is to find the empire of the Great Paitití—"that New World," a place of fabled wealth in southwestern Amazonia, and home to many potential new converts to the faith.[2] Don Benito has pushed through rainforests and across rivers to the borders of the Paitití's supposed domains. As evidence of his accomplishments, he has brought back to Potosí four or five Indians from the forests. These now stand in the church ready to be baptized. Their godparents will be Antonio López de Quiroga and his close relatives. The ceremony is performed, with "thanks to our Lord that, through the industry and expenditure of the said *maestre de campo*, these souls have been secured for Heaven. . . ."³

At about the same time that this service of baptism took place in Potosí, Antonio López embarked on what would be his proudest entrepreneurial achievement. Two hundred miles southwest of Potosí lay a small mining camp called San Antonio del Nuevo Mundo, a speck of habitation on the semidesert of the high Andean plain. Silver ores had been found here in the 1640s, and San Antonio had flared briefly and rowdily as they were mined and refined. But when the miners dug deeper, water rose in their excavations and prosperity ebbed. López, however, smelled profit. If a drainage tunnel could be cut to meet the waterlogged mines, the problem would be solved with one deft stroke. He negotiated with the mine owners. They agreed to give him a half share in their workings if his tunnel removed the water. Work began on December 5, 1672. At two o'clock on a winter's morning, almost five years later, a cascade of water into the tunnel marked the triumph of the venture. Spanish overseers and Indian laborers had blasted through half a mile of hard rock toward the veins of silver ore, and then had struck upward with a forty yard chimney. A final detonation of a cartridge now blew an opening into the lowest level of a working in the Rich Vein of San Antonio del Nuevo Mundo. By 3 A.M. all the water had drained from that vein. San Antonio was at work again, and Antonio López de Quiroga advanced in wealth and fame.⁴

Another fifteen years pass. Antonio López, now in the final decade of his life, still dominates San Antonio del Nuevo Mundo. And because of him, the former governor, or *corregidor*, of the town is an indignant prisoner in the town hall of La Plata. The governor, General don Gregorio Azañón y Velasco, has done his duty with zeal and courage, and all that his efforts have brought him is confinement. As if that were not enough, he is enmeshed in a legal and political web from which there seems to be no escape. He has no doubt who is at the center of this web—Antonio López de Quiroga. Having failed to secure redress from the viceroy of Peru, Azañón now writes, in offended but carefully tempered tones, to the Council of the Indies, laying out the events of the past months.

Soon after taking office, he reports, he tried to stop miners at San Antonio from smuggling the silver they produced out of South America through Buenos Aires. The law requires that silver from the mines around Potosí should be taken there for assay and for payment of a royalty of a fifth. But from places in the south of the district, such as San Antonio, it is easy and far more profitable to send the silver across the plains to Buenos Aires without paying the tax, and there use it to buy contraband goods from Brazil or Europe. As a good servant of the king, Azañón explains, he attacked this tax evasion. In doing so, alas for him, he ran headlong into Antonio López and his influential son-in-law, Captain don Miguel de Gambarte. He seized 50,000 pesos of theirs bound for Buenos Aires, whereupon López—"a powerful man who has held my predecessors almost at his command and disposition, on account of the mining interests he has in that province"—urged other miners at San Antonio to lodge complaints against their *corregidor* in the high court of La Plata. Next, Azañón recounts, López pulled out all the workers from the great drainage tunnel he had made, in an act calculated to bring mining at San Antonio to a rapid halt. And then he and Gambarte wrote threatening letters to Azañón. Finally, López persuaded the high court to send one of its judges to investigate matters in San Antonio. This was the Licentiate don Diego Reinoso y Mendoza, a man "whose violent spirit made me suspect how he would behave," writes Azañón, "even before he arrived." These fears proved only too well founded. Reinoso formed an alliance with the Quiroga faction in San Antonio, which Gambarte's nephew, don Joseph de Ujúe y Gambarte, commanded. Azañón rallied a few faithful officials and supporters among the miners. Dramatic confrontations, insults, and challenges in the plaza followed; nocturnal plots; rumors of riot; displays of arms; swordplay in the cemetery; a dawn raid on the miners' settlement outside the town; houses set afire there; a supporter of Azañón roused from peaceful sleep with his wife, and shot dead, point blank, by Ujúe y Gambarte; Azañón's goods seized and sold, and his wife's too, by Reinoso.

Being so outnumbered, Azañón explains, he then left for Lima to
beg the viceroy for support and justice. But the judge Reinoso
rushed in an adverse report that reached Lima before Azañón. The
viceroy ordered Reinoso to renew his investigation into the original charges, appointed another *corregidor* to San Antonio, and
put Azañón in jail. Matters improved, but briefly. From prison,
Azañón sent the viceroy his report on the affair, and was released—
but only to be ordered back to La Plata to seek justice in the very
court of which Reinoso was a member. His petitions have led to
a new confinement, in the town hall of La Plata; and from it, in
October, 1692, Azañón now complains to the Council of the
Indies, the ultimate court of appeal in the empire, that the attor-
neys of the town, "fearful of bringing charges against the said
judge," refuse to defend him. He has saved the crown thousands
of pesos in lost silver taxes. His reward is to be imprisoned on a
thousand pesos' bail, to have lost his office and his salary, and to
have seen many of his and his wife's goods seized and sold. How
much more comfortable it would have been, he must have thought
as he laid out his tale of woe for the Council of the Indies, to
have accepted Antonio López's mastery in San Antonio del Nuevo
Mundo, and slipped into his pocket, as earlier *corregidores* there
had done.[5]

In the records of Antonio López's life that remain for the historian to see, the hints of easiness and forgiveness in his conduct
are few and far between; while the hints of will, determination,
and even a certain implacability are frequent enough. Most evident of all his attributes is wealth. The picture has consistency.
From his earliest days in Potosí his no-nonsense qualities are clear
enough. They, though obviously not they alone, helped bring him
a fortune. As his wealth increased so, too, did his sense of right-
ness, turning, perhaps, an authoritative personality into one that
was increasingly authoritarian. Money, also, he could and did use
to gain political power, which then served to consolidate and
advance his material gains, while rendering him ever less toler-
ant of opposition. Only great age, perhaps, was finally capable of
softening him. It was not always a pretty life. But is is one that
cannot be ignored if the history of Potosí is to be properly told.

Gaspar Miguel Berrio, *Descripcion de zerro rico e ymperial villa de Potosi* (1758)

I

BEGINNING

TO POTOSÍ

BY ANTONIO LÓPEZ DE QUIROGAS OWN STATEMENT HE ARRIVED IN the Villa Imperial of Potosí late in 1648.[1] The first surviving proof of his presence there is the record of a loan of 1,400 pesos he made in April 1649 to a man named Felipe García de Alcántara. The document is a bare notarial transaction. It does not say what the recipient's business was, nor why López advanced him money (except to comment that the loan was a matter of "friendship and good works," which was standard notarial jargon).[2] It signals, nevertheless, the beginning of the career of the most remarkable and diverse businessman—moneylender, merchant, landowner, stockraiser, political manipulator, and above all miner and refiner of silver ore—that the extraordinary and Imperial Town of Potosí had seen in its 104 years of existence. That career lasted fifty years, almost to the month. López died in January 1699, having become a personification of the opulence and energy of what, even in decline, was the preeminent mining center of the Spanish American empire.[3]

The course of Antonio López's life before his arrival in Potosí is obscure. References in his own writing to his origins are few, general, and conventional (in that they are mostly allusions, offered as part of some petition for preferment, to this or that distinguished ancestor). Clearly he, at least, considered what he had done before 1648 as inconsequential. There is good reason to think that he was in Lima in the early 1640s (he was certainly there by

7

January 1642), and before that in Seville. In both cities he was in contact with merchants in what he called the "carrera de galeones"—that is, men shipping goods from Spain to the South American colonies, and other items, mainly silver, back again.[4] Whether he himself went in for this trade is open to question. If he did, he makes no mention of the fact later; on the other hand, once in Potosí he showed a fine flair for handling money, which perhaps came from some previous experience with buying and selling.

Only an adult would have been familiar with the business of the transatlantic shippers in Seville, who were rich and powerful men. So Antonio López in Seville around 1640 must have reached, if not the full twenty-five years required in Spain for legal majority, at least something close to that age. Perhaps, then, he was born around 1620. No record of his birthdate has yet come to light. But if indeed that year is near the mark, then it is reasonable that the renowned chronicler of Potosí, Bartolomé Arzáns de Orsúa y Vela, should have described him as very old in his last days in the late 1690s. "His life was long, and although he raised his food to his mouth with difficulty, he permitted no other hand to help him."[5]

At least the place of López's birth is definite: the parish of Triacastela, in the bishopric of Lugo in northwestern Spain.[6] (See Preface, above.) Lugo is part of the larger province of Galicia—the green and damp upper corner of the Iberian peninsula that projects out into the Atlantic. The village of Triacastela, however, is almost as far from the coast as any place in Galicia. It is set in one of the many deep valleys that lie among the rough ranges dominating the country to the south of Lugo. The sides of these valleys are covered with dense coppices of chestnut, which until recent times provided country-dwelling Galicians with one of their staple foods. Interspersed with the trees are vineyards, the source of the thick red, and acid white, wines of the region. The valley bottoms offer good land for raising grain and root crops, and grazing cattle. Streams draining from the mountains give fish. Squat, square farm houses with grey stone walls and shallow-pitched roofs of slate crouch on the slopes. The scale of these valleys is intimate, the impression one of abundance, and the views full of the beauty that comes when a well-favored land is carefully tended by its people. A greater contrast in scale and scene to Potosí cannot easily be imagined, though if Antonio López ever missed the green valley of his youth, he never mentioned it.

Quirogas had lived in this corner of Galicia for many centuries. Indeed, not far to the south of Triacastela, over the mountains, are the castle, now scarcely even a ruin, and the town of Quiroga, set in another valley of imposing beauty.[7] The historical record takes the family back to the early thirteenth century there. But tradition holds that it was a Quiroga who, using the dense woodlands of the valley for cover, checked the entry of the Moors into Galicia in 715. His main defense consisted of stakes tipped with sharpened iron; and in imitation of this, the family later took as its heraldic arms five silver stakes on a field of green.[8] It is not recorded whether the Quirogas ever worked the local iron deposits; but Antonio López de Quiroga certainly did justice to the silver stakes with his mining activities in Potosí.

In a description of Galicia compiled in 1550, the name Quiroga appears as an "apellido de casas solariegas" (a surname associated with an ancestral seat), and the family is ranked as being of "honrados hidalgos" (honorable noblemen).[9] They were clearly, though, no more than minor provincial nobility, of far lesser standing than, say, their near neighbors to the southwest at Monforte, the Counts of Lemos. No branch of the family ever acquired a title; and when in later life Antonio López sought one bearing the name of his American estates, his requests were consistently rejected. The line occasionally, nonetheless, produced men of great talent and energy. One of them was Vasco de Quiroga, a member of the reforming second Audiencia, or high court, of Mexico in the early 1530s, and still better known for his later efforts, as first Bishop of Michoacán, to protect and organize the Indians of western Mexico. Another, whom Antonio López liked to point to when requesting rewards for his own services to the Spanish crown, was Gaspar Rodríguez de Quiroga, Cardinal-Archbishop of Toledo in the late sixteenth century, Inquisitor General and Grand Chancellor of Spain, President of the Councils of Italy and of State under Philip II.[10]

Antonio López de Quiroga was the product of a junior branch of the family. His father, Alvaro de Quiroga y Rivera, was a younger brother of the lord of the town of Valdefarina, Pedro López de Quiroga y Rivera, who had an established residence in Triacastela. Another brother, and therefore uncle to Antonio López, was don Juan de Quiroga, who became a knight of Santiago and Captain of the Guard to don Francisco Ruiz de Castro y Portugal, eighth Count of Lemos, appointed Viceroy of Sicily in 1616. Some sort of permanent relationship, though with strong servant-to-master overtones, seems likely to have developed between at least this branch of the Quirogas and the Counts of Lemos. Antonio López was certainly known to, and perhaps on familiar terms with, one of the Counts: don Pedro Fernández y de Andrade, Viceroy of Peru from 1666 to his death in 1672. The branch of the Quiroga family to which López belonged had its origins near Monforte de Lemos.[11] And López's mother, doña María Fernández de Reimóndez, was from a place not far away from there—Sarria, ten miles west of Triacastela, and eighteen north of Monforte. Reimóndez (or Raimóndez) was another old Galician name, of equal rank with that of Quiroga.[12] Antonio López, however, never included either of his mother's surnames in his own, and rarely made any reference to her family in the various reports on his services that he submitted to the crown in the prime of his life.[13]

Triacastela today, like the region around it, is remote and little visited, but it was not always so. For the village lay on the pilgrims' road to Santiago de Compostela, the site of the shrine of Saint James the Great. In medieval times, when the pilgrimage was at its most active, many foreigners passed along this way en route to Spain's most venerated shrine. This cosmopolitan procession can only have stimulated the curiosity of the town's more venturesome inhabitants. And so, although by Antonio López's time fewer travelled to worship at the Apostle's supposed tomb than in earlier centuries, it is no surprise to find Quirogas leaving their enclosed valley, and spreading across many of Spain's territories.

As the example of Vasco de Quiroga shows, Antonio López was far from being the first of his family to make the transatlantic journey. He must have known this, although he never alludes to predecessors from his family in the colonies. Most of them were associated with Chile, and among these stood preeminent a member of the second branch of the Quirogas, don Rodrigo López de Quiroga y Sober (1512–1580), who was among the founders of Santiago de Chile with Pedro de Valdivia in 1541, and then held a series of civic and administrative posts in the colony, rising to become its Governor and Captain General in 1575.[14] He is reported to have fought the Araucanian Indians of southern Chile, and also Sir Francis Drake (presumably while Drake was rampaging up the coast of Chile at the end of 1578, in the early stages of his famous circumnavigation of the globe).[15] In the conflict with Drake a nephew accompanied don Rodrigo, by name don Antonio de Quiroga y Losada.[16] Here comes into view another family that is inseparably linked with the life of Antonio López de Quiroga, as indeed it had been with many generations of Quirogas. The Losadas were fellow Galicians, residents indeed of the valley of Quiroga itself, where they had lived since at least the twelfth century. Over the succeeding centuries the two lines had grown completely intertwined.[17] Like the Quirogas, the Losadas prospered in Chile in the sixteenth and seventeenth centuries. A notable representative of the family there was Antonio de Losada y Villasur, lord of Cubillos (in the kingdom of León, near Ponferrada, and some 40 miles east of Quiroga), who moved to Chile in 1575 as a young man. There in 1579 he married, in Santiago, a relative, doña Inés de Gamboa Quiroga.[18] It seems at least likely that some of the people bearing the Losada and Quiroga names who appear during the seventeenth century in and around Potosí had their origins in Chile; though others, like Antonio López de Quiroga himself, came straight from Spain.

There are indeed several Quirogas to be seen in seventeenth-century Potosí, and the province of Charcas surrounding it, before Antonio López himself makes an appearance. In fact, in September 1620 no less than another Antonio López de Quiroga was in the town, receiving a power of attorney from one of the public notaries and an associate to prospect for silver in the newly discovered ore deposits of Macha, and also witnessing a will.[19]

Despite the identity of names, this can hardly be the great Antonio López of the second half of the century, since such an early appearance as an adult would make him over 105 at his death in 1699. Still, it is certainly suggestive to see that this first Antonio López apparently had some interest or ability in mining, and perhaps not too far-fetched to imagine him relaying reports on mining in Charcas back to Spain. Perhaps this López remained some years in the district, for in 1636 an Antonio López de Quiroga again appears in the Potosí area, this time as a financial guarantor of the *corregidor*, or town governor, of the mining center of Porco, a few miles southwest of the Villa Imperial.[20]

Yet another Quiroga engaged in mining in the mid-1630s was Captain don Andrés Somoza Losada y Quiroga, owner of silver mines in the province of Carangas, north of Potosí.[21] The combination of surnames suggests that this Quiroga was close in family origins to Antonio López de Quiroga, who once appended Somoza to his own name (see n.13 above). It also definitely raises the possibility of a close link between don Andrés and still another Quiroga not far from Potosí in the 1630s and 1640s—a man, moreover, with whom Antonio López quite unequivocally had dealings. This was the governor of Santa Cruz de la Sierra, don Juan Somoza Losada y Quiroga. The use of identical surnames might suggest that don Juan and don Andrés were brothers; but neither mentions that fact in the evidence that has so far appeared, and the liberty that Spaniards took with the use of names makes it difficult to be sure of relationships even in such an apparently obvious case.[22]

Don Juan Somoza Losada y Quiroga, knight of Santiago, was appointed to the governorship of the province of Santa Cruz de la Sierra, in the eastern lowlands of Charcas, in 1637.[23] He was Galician by birth, from the bishopric of Lugo, and the son of Alvaro de Taboada y Quiroga and doña María López Vizcaína. (Taboada is a town about thirty miles west of Triacastela, although don Juan gave his own birthplace as Laxosa, a small place just east of Lugo.) Early in the 1640s he was given another governorship in South America when the Marquis of Mancera, the ruling viceroy of Peru, named him to Tucumán, in the north of what is today

Argentina. He seems not to have taken up that office, however, and may have returned at some later date to Santa Cruz. But it is through this appointment that his connection with Antonio López de Quiroga becomes apparent, because López presented the viceroy, while the appointment was still under consideration, with a memorandum (*memorial*) from don Juan. What precisely this document contained is not reported, but it was probably simply a report on services performed that don Juan himself had submitted. Antonio López took the paper to the viceroy as don Juan's legally appointed agent in Lima, and Mancera included it in the *decisión* of appointment issued to don Juan of January 5, 1642.[24] Here, then, are the first two pieces of concrete information about Antonio López de Quiroga in South America: he was present in Lima by early January of 1642; and he had a relationship of some trust by that time with a prominent man who bore his family surnames.

No hint comes from the documents, alas, about what Antonio López was doing between that appearance in Lima and his arrival in Potosí late in 1648. The best guess, which can rest only on his declared acquaintance with international merchants in Seville and Lima in and after 1640, is that he occupied himself in some commercial capacity in the mid-forties. Possibly it was this work that provided him with the money that he began to lend soon after getting to Potosí. Some earlier experience in trade would also fit with Arzáns's assertion that López set up shop in the Calle de los Mercaderes, or Merchants' Street, just off the central plaza, soon after his arrival. In this he was helped by "a few good men," whom Arzáns, all too characteristically, does not identify.[25] Logically these would have been Galicians or people with whom López had had commercial contact from Lima. But neither his, nor anyone else's, statements shed any light on this question.

On the other hand, it is perfectly clear that there was one particular family in the Villa Imperial from which López received much help in his first few years in the community: the Bóvedas. From them, indeed, he took not only material aid, but also a wife. The head of this prosperous household was Lorenzo de Bóveda, husband of doña Ana María Bravo de Saravia. A Galician connection is certain here, for the Bóvedas were "an illustrious and ancient house of this kingdom of Galicia,"[26] and a family, furthermore, with links in the remote past to the Somozas. Their seat in Galicia was a small town of the same name, a few miles north of the home of the Counts of Lemos at Monforte, and only sixteen miles southwest of Triacastela. Once more, little imagination is needed to picture Antonio López's being oriented, and perhaps welcomed, in Potosí, not just by fellow Galicians, but by Galicians from precisely his home ground. And, although no Quirogas may have been at hand to take him in when he arrived, the acceptance he obviously found can only have been increased by the earlier presence in and around Potosí of other members of the Galician families from which he descended. Furthermore, a newcomer carrying the trust, and two of the surnames, of a current, or recent, governor of Santa Cruz de la Sierra, can only have had a definite advantage in Potosí over the raw immigrant from Spain.

The Bóvedas were prominent in the town by the time Antonio López arrived. Lorenzo had by then acquired the honorary military title of *maestre de campo* (roughly equivalent to the modern rank of colonel), a sign of rapid rise to distinction, since a few years before he had held neither of the two lower ranks of *sargento mayor* or *capitán*. Such a title might be bought or be awarded as a mark of official esteem. In either case, to have it denoted success in the public view. Furthermore, Bóveda and his wife were regarded as people of nobility (*personas nobles*) in the Villa Imperial. Though that can have been only a local ranking (in Galicia, the name Bóveda did not match Quiroga, Somoza, or Losada in eminence), nevertheless it must have drawn Antonio López to the family.[27]

Behind these marks of social distinction lay rising wealth. Lorenzo de Bóveda begins to appear in notarial records by 1640 as a merchant dealing in imported and local goods. (A reference from that year, indeed, shows him despatching iron, other unspecified Castilian goods, and salt from Potosí to San Lorenzo de la Barranca in the province of Santa Cruz de la Sierra, of which Somoza Losada y Quiroga was by that time governor—though any connection between the two men must be conjectural).[28] By the late 1640s he was regarded as a substantial merchant (*mercader grueso*) in Potosí. Some confirmation of this status can be taken from the fact that in a mere two days in October 1648, he made three loans worth, in total, slightly over 38,000 pesos—a sum that would have bought five or more very substantial houses, or nearly 10,000 llamas.[29] The use to which this credit was to be put is not recorded in the loan notes, but a later reference strongly sugggests that Bóveda advanced credit to miners—a practice to be picked up in due course by his future son-in-law.[30]

Lorenzo de Bóveda died in the middle of 1652. By that time, Antonio López had married into this large and prosperous family (his bride, doña Felipa, was one of at least six children—two daughters and four sons—borne by doña Ana María). The confidence that the family had in their new relative is shown by the fact that López became the executor of the estate. If Arzáns is to be believed, he also inherited most of the family's wealth, which he then used as the foundation on which to build a far larger fortune.[31] But López never alludes to any inherited cash or business received from his father-in-law—an omission that may reflect either reality or an unwillingness to acknowledge that he had had any such help in the initial successes of his career.

POTOSÍ

Whatever the truth about the inheritance (and even if it had not been as large as might be desired, doña Felipa must at least have received a share, besides having brought a dowry with her), López de Quiroga can only have felt gratified as he contemplated his position after his first three or four years in Potosí. There he was, a man of probably little more than thirty, well ensconced in the society of a town of fabled wealth, newly married, and the executor of the estate of a prosperous merchant. Much, perhaps all, must have seemed possible to him. And indeed there is no negative note to be found among the manuscripts that document his activities at this stage of his life.

But Potosí. . . . As he surveyed the Villa Imperial and especially its economic affairs, López certainly saw that all was not as reputation in the world at large proclaimed. Wealth there still was in abundance, and Potosí was indeed the center of a region that produced more silver than any other mining district on earth, and remained the envy of Spain's rivals and enemies. But the glitter on the fable was tarnished, and seemed likely to be dimmed entirely if things continued as they had been going.

The peak of prosperity was indeed long past. It had been distant when Antonio López was born. The days of great glory, of Potosí's emergence from the remote Andean wastes as a wondrous engine of Spanish power, lay far back in the final thirty years of the sixteenth century. The all-time maximum of silver output had, in reality, been reached in 1592, when ores wrested from the fabled Rich Hill (Cerro Rico) on the southern edge of the town had yielded 444,000 pounds of fine silver. By the standards of the day, this was an enormous quantity, equivalent in its value to nearly 44 percent of the total annual expenditures of the Spanish crown in Spain and the rest of Europe in the mid-1590s.[32] But from the early nineties the trend of production was inexorably and evenly downward (see fig. 1). True, there was no dramatic slump. Potosí's output did not collapse as abruptly as it had risen from the mid-1570s to the mid-1580s, propelled upward by the twin stimuli of cheap forced labor and new methods of processing ore with mercury.[33] The dramatic surge in production that had occupied scarcely fifteen years took a hundred and forty to subside. Only around 1710 did the amount of silver being taxed in the Potosí treasury finally drop to the levels typical of the early 1570s. But, with occasional exceptions when new discoveries of ore out in the district provided relief (as in the late 1630s, when a place called Chocaya briefly flourished), the movement of production and prosperity in the Villa Imperial was downward throughout the seventeenth century.

What had happened? The best explanation seems to lie in the geology of the Rich Hill. When the Spaniards chanced on this outwardly unremarkable formation in 1545—just one minor peak among thousands in the eastern ranges of the Andes—they rapidly realized that here was an unparalleled geological fluke. Four

Annual Registered Silver Production in the Potosí District, 1550–1710

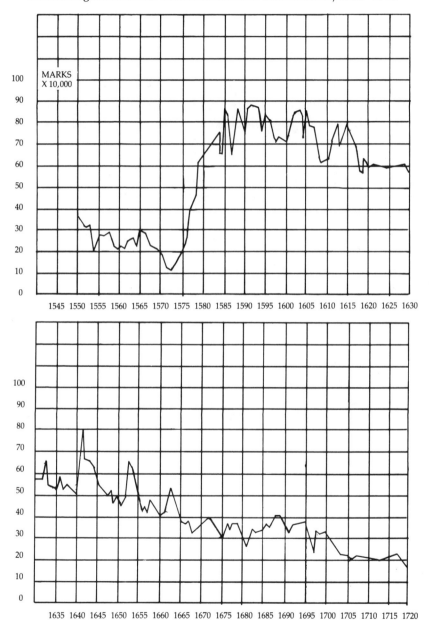

For sources see: Peter Bakewell, "Registered silver production in the Potosí district 1550–1710," Jahrbuch für Geschichte von Staat, Wirtschaft und Gesellschaft Lateinamerikas, 12 (1975), pp, 67–103.

thick veins of rich silver ore lay packed tightly together in the top few hundred feet of the Hill. Not only were the veins large and rich, but they were simple to attack. The main difficulty was posed by height, since the summit stood at about 15,700 feet. But this was no more than 2,600 feet above the level of the town of Potosí, just a few hours climb from the main square, and not demanding a great physical effort from native Andean workers, with their inborn physiological adaptation to altitude, or even from Europeans, once they had lived in the highlands for a short while.[34] Certainly Potosí was a cold place to work, and the top of the Hill even more so. But the climate is at least dry, so that storms and snow were an infrequent hazard. This scarcity of rainfall also had the advantage of reducing the amount of groundwater in the Hill—at least in its upper reaches, where the ores lay most tightly packed. When mining extended to lower parts of the veins, water and flooding certainly became a problem. That, though, was a difficulty that seventeenth century miners had to deal with, and prominent among them in doing so was Antonio López. In the sixteenth century, miners of the silver ores of Potosí, by contrast, had an almost ideally simple task to perform. Astonishingly rich ores were crowded into the steeply conical summit of the Hill. Given the shape of the formation, the veins could not be far from the surface. Little underground work was necessary to tap them; flooding was not a problem; and, as a final advantage, the rock in which the veins lay was firm, so that tunnels needed little reinforcement. The result of all this was that great quantities of rich ore were rapidly extracted, at low cost, in the sixteenth century. When, in the 1570s, cheap labor, and a new refining process capable of handling the ores in great bulk, were added to the geological situation, an unprecedented boom of silver production occurred.

But the same geological arrangement that brought the boom also, inevitably, brought decline. The very ease with which the best ores in the Rich Hill, concentrated in the summit, could be extracted meant that they were rapidly depleted. Even before the sixteenth century ended, indeed, miners in Potosí were having to dig deeper, and go lower down the Hill, to find good ores. And as they did so, again because of the conical shape of the Hill, the distances they had to dig in order to reach the veins tended to increase. So their costs rose. It was not that the Hill lacked good ores in its lower reaches. On the contrary, high grade deposits lay toward the base. There, however, the volume of rock to be searched before those deposits could be found was geometrically greater than in the summit. And water was a constant foe in the lower levels. Mining there proved to be a dogged game of long-term and uncertain investment in exploration and costly drainage works. Antonio López de Quiroga became the best player of that game in colonial times.

Increasing inaccessibility of ore, then—translating itself in practice into higher costs of extraction—set Potosí's silver output on its descending track after the 1590s. Another source of higher costs, clearly apparent by the end of the century, was Indian labor. In the 1570s a complex system of native forced labor, known as the *mita*, had been set up by the viceroy of the day, don Francisco de Toledo, to supply workers cheaply to the mines and refineries of the town. By 1600, that system, even though it was designed to provide Potosí with between four and five thousand active laborers at any particular moment in the year, was clearly falling very far short of supplying all the manpower that was needed. The labor requirements of silver production had grown far above the ability of the Indian communities to meet as prescribed in the original *mitas* of the 1570s. Moreover, decline in the Indian population through disease and maltreatment, coupled with migration out from the communities included in the *mita* system, meant that ever fewer native workers actually arrived in Potosí in the drafts. So miners and refiners had turned to Indian wage labor to fill the gap, to such a degree that by 1600 slightly over half the Indian workforce producing silver in Potosí seems to have been contracted wage labor.[35] This was a good solution to the problem of labor, in the sense that it seems to have prevented an absolute shortage of workers from arising. Silver producers in Potosí never complained that they simply could not find men to mine and refine ore. But the solution was expensive, because enough wage workers could be secured only by paying them more than was given to the drafted men. In reality, of course, the wages paid to the salaried workers were determined by supply and demand. Around 1600, when demand for manpower was still high, a wage laborer engaged in cutting ore in a mine earned three to three and a half times more than a draftee working below ground. By the 1630s, when demand for labor seems likely to have declined somewhat, the difference had narrowed a little.[36]

So the increasing costs faced by silver producers in the Villa Imperial by the end of the sixteenth century had two components: extraction of ore and labor. It is difficult, however, to see anything else in the circumstances of the mining industry that changed significantly, let alone adversely, at that time. On the face of it, mining seemed very well set by 1600. Techniques were fully established, with constant small improvements being made in the efficiency of the refining process with mercury that had taken root in Potosí in the 1570s.[37] And though practices of ore extraction were primitive, they certainly had not deteriorated during the boom years. If anything, in fact, the number of skilled miners, capable of intuitive "reading" of workings for good ores, had grown.

In its physical structure, too, the silver industry was fully equipped by the beginning of the new century. The most impressive part of this structure was a large number of refining mills in or near Potosí—seventy-two powered by water and six driven by horses or mules, according to an anonymous writer of 1603.[38] Also crucial for producing silver were a series of dams that stored water to drive the ore-crushing machines in the refineries. By 1620 there seem to have been about thirty of these, built across narrow glacial valleys in the high Kari Kari range just a little east of Potosí.[39] The water was led down through aqueducts and tunnels from the upper reservoirs to the lower, and eventually into a canalized river bed, called the *Ribera*, that dropped down through the town. Along each side of the *Ribera* the refineries were ranged, each with its own lesser canal drawing water off to power a vertical wheel, and then returning it to the main channel for use in the refineries downstream. This was a system built up mainly during the boom years of the 1570s and 1580s, to make it possible to process ores the whole year round. The stream water flowing naturally through Potosí had not yielded enough power for milling during the dry months of the late autumn, winter, and early spring. But with what the dams could store, milling, by 1600, was possible during the entire year, if rainfall was normal. The whole structure of dams, tunnels, and canals represented an enormous investment of cash and labor, which stood ready to benefit silver producers in the seventeenth century, at no cost except that of upkeep.

In addition to permanent physical plant, of course, seventeenth century miners and refiners needed constant fresh supplies of raw materials. But getting these supplies seems to have caused them no more than passing difficulties. The list was long, silver-making being a complicated business that demanded many and varied ingredients: wood for machinery, buildings, and fuel; iron, also for machinery, and, in the form of filings, as a reagent in refining with mercury; steel, for tools; salt, as a reagent in refining; llamas and mules, for freight of ores down from the Hill to the mills below, and of goods of all sorts into Potosí; and, most crucial of all, mercury—the key to refining. Despite the number and variety of these items, though, no evidence suggests that any of them was in permanently short supply in the seventeenth century.[40]

Anyone wanting to mine or refine silver in Potosí in the early 1600s, then, would seem to have had much in his favor. But all the advantages were negated by one large and fundamental hindrance: the depleted state of the rich veins in the upper reaches of the Rich Hill. That silken purse, once crammed with silver, had become a sow's ear full of rubbish, and there was no reversing the change. What needed to be done now was to explore the Hill for other concentrations of wealth. Seventeenth century miners proceeded to do that. But it was a large, slow, and costly business; and it was not until Antonio López threw himself into it that anyone proved wholly successful at the job.

For all that Potosí's economic base began slowly to crumble after 1600, this was nonetheless still a place of dazzling allure for many in the Spanish world (and outside it)—an allure that draws the modern historian as well, though none has yet taken the full measure of the town and its people.

They were so many people. In size of population, if in no other respect, Potosí in the early 1600s was an Amsterdam, a London, or perhaps even a Seville or a Venice, laid on a dry, tawny upland of the eastern Andes. As early as 1561, before the great boom in mining began, 20,000 Indians were reported living in Potosí, nearly all of them—19,700 to be precise—making a living from something else besides silver: "unimportant tasks such as making candles and bread, and selling fruit and other things to eat," said a rather severe official visitor to the town, Juan de Matienzo, in that year, thinking that more than three hundred of them should be out in pursuit of silver.[41] The people being counted here seem to be men only, though it is quite clear from other sources that many native women and children were also in Potosí by this time.

Many more people were drawn by the expansion of mining in the 1570s. "New people arrive hour by hour, attracted by the smell of silver," wrote Matienzo in 1577, once again in Potosí on official business. The town now had 2,000 Spanish heads of household (*vecinos*), a "great multitude of [Spanish] women and children," and 20,000 or more Indian men, with an equal number of women and children.[42]

Twenty-five or so years later, just after the peak of the boom, the number of native males had at least doubled, according to estimates of various people familiar with the Villa Imperial. One apparently reasonable calculation (although the writer confesses finally that nobody knows for certain how many Indians live in Potosí) puts the number of native men at about 50,000 at the turn of the century: 12,600 draft workers there for the *mita*, 10,000 skilled men working for Spaniards, 8 to 10,000 passing through, and 20,000 who "occupy themselves from day to day."[43] Of all these, only 12 to 13,000 did tasks directly related to mining at any one time, so that many others must still have been spending their days on "unimportant" labors—building, smithing, hat making, tailoring, carpentering, and other crafts; preparing and selling food; spinning and weaving; cleaning and carrying; tending llamas, alpacas, sheep, pigs, horses, mules; searching out and cutting firewood; making charcoal; freighting fuel, wheat, maize, potatoes, *chuño*, *quinoa*, beans, wine, mercury, salt, iron, stone, and silver. The 1,500 Spanish residences, large and small, reported to be standing at about the same time must have occupied many Indians as servants. These houses were inhabited by over 3,000 Spanish men.[44] The adult male population, adding Indians to Spaniards, exceeded 50,000 (to take no account of people of mixed race, free and slave blacks, and sprinkling of foreigners—Portuguese, Italians, and an occasional northern European). Since many of these were married, and had children, the total number of people, young and old, in Potosí by 1600 was very probably over 100,000.[45] After then, it is hard to say for sure what direction the population took. There are fewer reports than before from visitors and officials about numbers, which suggests that no dramatic changes occurred in the new century. A gradual decline, in line with the downwar drift of silver production, seems the most likely trend. Nevertheless, the population was still large when Antonio López appeared, half way through that century. Potosí was still one of the great urban centers of the Spanish Indies, to be measured only against the viceregal capitals, Lima and Mexico City.[46]

This great mass of people did not sit easily together in the town. Life in Potosí was strenuous, turbulent, often violent. Even merely getting to the town demanded effort. At over 13,000 feet, set amidst the eastern range of the Andes, with a hundred miles of altiplano and then a second chain of imposing mountains separating it from the Pacific, Potosí was one of the least accessible towns in Spanish America. Only the determined, the ambitious, the avaricious, the very needy, or, in the case of Indians going to the *mita*, the compelled, completed the journey. And the newcomer, once arrived, was cast into a microcosm, insulated by space, whose workings rested to a large extent on the operations of chance. With so little known of geology, and so few means of subterranean exploration available, mining was inevitably at best a matter of intuition, and at worst of outright luck. Little wonder, then, that Arzáns de Orsúa y Vela, Potosí's prime chronicler, should exclaim so eloquently at the fickleness of fate and the fragility of life: ". . . a humor that rises to the head. . . , a passion that occupies the heart. . . , a tile that falls from above. . . , a mistake in counting. . . , a yawning open of the earth that swallows one up, [these] and a hundred thousand other events open the door to death and are his ministers."[47] But little wonder, also, that in this atmosphere of unpredictability and changeability, those who had braved obstacles to reach Potosí tended to intemperate, arrogant, defiant action. If Potosí was larger than life in the mind of the outside world, so too were its inhabitants inclined to extravagant habits and overweening gestures. Miners, and especially miners of precious metals, are given, in the popular view, to raucous living, to drinking, gambling, and womanizing. To this expectable boisterousness was added in Potosí awareness that here, even in decline, was the greatest source of silver known in the world, indeed the greatest the world had ever known. The outcome was a ferocious grandiloquence palpable in many aspects of the town's life.[48]

Outsiders were impressed with all aspects of the place. So great were the profits to be made on trade, observed one visitor from Spain in the early seventeenth century, "that no sooner have figs and grapes ripened than they appear in the plaza—before the viceroy has them on his table in Lima." An early fig, or an egg, cost an entire silver *real*.[49] Pack trains of five hundred llamas carried goods into the town, and no fewer than a hundred thousand of these beasts were needed to keep the inhabitants fully supplied. One major cargo was wine: Spaniards and Indians together drank more than 150,000 jars (*botijas*) of it every year, in addition to which the Indians spent a million pesos a year on their corn beer (*chicha*), of which they imbibed so much on Corpus Christi day alone that "many rivulets of urine can be seen running in the streets from this remarkable gathering of people."[50] The Spaniards took their wine, if not the Indians their *chicha*, from silver vessels, for much of Potosí's output went into making plates and containers, and "in all Peru it is only the very moderate man who does not have silver tableware."[51] Antonio López doubtless had this sort of comment in mind when in 1676 he included over 7,000 pesos worth of silverware—cutlery, plates, jugs, basins, salt cellars, and the like—in the dowry carried by his daughter, doña Lorenza, to her marriage with the *maestre de campo* don Juan de Velasco; though in truth this was a rather insignificant part of the 100,000 pesos that the dowry was worth as a whole. Twenty sacks of coin, containing 49,075 pesos and 2 reales, about a ton and a half of silver, were the single largest item bestowed on the new couple.[52]

But if the scale of public economic activity was large in Potosí, and private wealth might assume grand proportions, commensurate with these was the violence spawned by inequality, ambition, and greed. One of the judges of the high court in La Plata, visiting Potosí in 1585, found the town to be a den of thieves, a Babylonian place—and its Spanish inhabitants "the most perverse sort of people the world has created." Writing more than a century later, Arzáns noted with distaste how often fights between individuals—the results of "very trivial causes, as was the custom in this town"—figured in his *History*[53]. But there is a pleasurable *frisson*, a certain furtive celebration, in his cataloguing of affrays, stabbings, poisonings, and vendettas, that betrays in him a feeling that violence was fitting in Potosí, part and parcel of its nature.

I shall only say that from the year of 1670 to this of 1702 . . . , in the houses, streets, plazas, and open spaces of this town, both in confrontations and in deeds of treachery, among many people or one against one, we may count sixty deaths (of those that are known) of Spaniards and Peruvians; and more than two hundred of *mestizos*, Indians, and dark skinned people. These, as I say, are the deaths we are aware of. God alone knows how many died secretly, for wherever trenches or foundations are dug in houses, stables, or open land, always are found the bones of those miserable beings who have perished at the hand of such bad Christians. . . .[54]

In the Potosí of Antonio López's century, the most severe episode of public violence was the gang warfare that flared up in the 1620s, between "Extremadurans and Biscayans, who have created much damage and disturbance over women, gambling, and other matters."[55] The roots of the conflict in reality were deeper than this, and the conflicting groups rather more extensive. Basques (*vascongados*) in general, rather than just those from the province of Biscay, formed one contending faction. The other included men from much of the rest of Spain, and also locally born inhabitants of Potosí. This second faction came to be known as *vicuñas*, apparently on account of the hats made from vicuña wool that they adopted. Behind the hostilities were certainly regional antagonisms brought from Spain, and especially the antipathy that many felt towards Basques—who by virtue of ancient legal privileges, possession of a different language, and a long history of distinct, if not separate, development in Spain, tended to cluster together in what others often considered an exclusive and disdainful way. Beyond that, in Potosí as elsewhere in Spanish America, Basques had risen to a position of dominance in silver production, perhaps helped by mining skills brought from the iron industry of their native provinces. From their prominent position in mining, they had progressed in the second decade of the seventeenth century to ever greater influence in municipal affairs, controlling in particular the two *alcaldías mayores*, or offices of first-instance magistrate—the most powerful positions in the town's administration, after that of the *corregidor*. Resentment toward them accumulated among those who were less rich and powerful, a resentment that can only have been exacerbated by the sharp fall in silver output that began in 1615 (see fig. 1). Many had been drawn to Potosí "by the smell of silver," as one observer of the initial troubles put it,[56] but now there was ever less to go round once the source of the scent was reached. Even the far off viceregal government in Lima began to foresee dangerous trouble in Potosí. In 1617 the viceroy of the time, the Prince of Esquilache, went so far as to annul that year's election of two Basques to the magistracies, hinting at manipulation of the vote in the town council.[57] Tensions still rose, however, and eventually, in June 1622, the murder of a prominent Basque on a Potosí street set in motion a wave of violence that lasted for two years. By March of 1624 the *vicuñas* had killed sixty-four men, and wounded countless others.[58] But they had come no closer to realizing their wish that Basques should be driven from the mines and town hall of Potosí, and their efforts began to flag. The *corregidor*, in alliance with the high court in La Plata, took advantage of this weakening to suppress the *vicuñas* by force; and finally a general pardon, granted by the viceroy, and covering all but the most culpable of the *vicuñas*, was proclaimed in April 1625.[59]

The gang war of the *vascongados* and *vicuñas* long remained a vivid memory in Potosí, and no less so in the minds of high colonial officials. On both sides, it had evoked an impressive ferocity. The brutality of both the participants' actions and of their words, recorded in contemporary documents, makes Arzáns's accounts of violent deeds in Potosí seem more plausible, less of a baroque hyperbole, than an initial reading of his *History* would suggest. In one incident, a wealthy Basque landowner was shot to death by *vicuñas* in a church doorway. His body was carried into the church, where one of the assailants drew his knife and stabbed the Basque twice in the face, for "they had said that they wished to have his blood on the dagger."[60] Viciousness of this sort did not soon fade from the official consciousness; nor was it easily forgotten that royal control in this key component of the Empire's economic structure had at least seemed gravely compromised by the factional mêlée. Small wonder, therefore, that when in 1641 an inspecting visitator of the high court of La Plata proposed reopening investigation of the affair, the viceroy Marquis of Mancera should have instantly ordered the court to gather up any findings that the visitator might already have collected, and lock them up with six keys. The inspector himself was to be sent directly back to Lima under the escort of one of the judges of the La Plata court.[61]

No Quirogas, Losadas or Somozas appear to have taken part in the war, at least not prominently enough to leave their names in the manuscript record. Galicians there certainly were, however, among the *vicuñas*. Some of these perhaps lived long enough, (since the *vicuñas* tended to be footloose young men) to witness the rise of López de Quiroga to eminence in Potosí, and perhaps to feel a tardy satisfaction as he, a Galician, built for himself a dominance in silver production far greater than that enjoyed by any one of the Basques early in the century—all the while gathering Galicians around him rather as Basques had clustered round their leaders in their heyday.

For all Potosí's brutality and turbulence, it was, nevertheless, far from being a raw mining camp. A European set down in the central plaza would have had no difficulty recognizing around him the features and signs of an organized town, though he would probably have missed the patrician elegance displayed by European communities with populations as big as Potosí's. "The houses of Spaniards are very good, by the standards of this part of the world," wrote the anonymous author of a well-known Description of Potosí of 1603, "but most of them have roofs of *ichu* straw."[62] Without a doubt, the visitor's eye would have been drawn first to the Rich Hill, overlooking the town a mile or two to the south—a natural object "of such lovely form that it seems made by hand," said the writer of 1603, adding that it looked like a large pile of wheat in color and shape, though its surface had grown rough with so much working.[63] Having taken in the form of this natural feature, and perhaps reflecting on the world-wide consequences of human exploitation of its interior—Potosí's silver being exchanged at Manila for Chinese silks, or being paid out to Spanish troops in the Netherlands as they strove to contain Protestantism—the visitor might have dropped his gaze again to survey the plaza. This was an impressive space, about 150 yards square, sloping moderately down from east to west, and surrounded by official and private buildings. Much of the public and private business of the town went on here. When, in 1614, one of the official notaries of Potosí rented out a part of his house on the plaza, he took care to stipulate in the contract that he reserved for his own occasional use a second-story window from which he could watch bull fights, and the processions and fiestas taking place during the week of the feast of the Holy Sacrament. (The only sort of window available to the poor renter was a hole in the wall covered with a piece of waxed cloth "in the room where the bed is.")[64]

Private houses and shops were outnumbered, however, around the plaza by public buildings. Which of these the inhabitants considered most important would no doubt have varied with their day-to-day preoccupations; but many would have pointed to the Iglesia Mayor, or principal parish church of the Villa Imperial, standing in the north-east corner of the square, and a building with manifold functions and meanings for *potosinos* of all stations in life. Here it was, for instance, that Antonio López and his family were to offer themselves as godparents of the Indians brought back by the expedition searching for the Gran Paitití. Here, also, fugitives from the town justices often took refuge. The Iglesia Mayor was very convenient for such purposes, for just fifty yards down the hill, in the northwest corner of the square, was the town jail. Escapees did not have far to run.

The jail, in fact, was merely one part of a large building on the west side of the plaza from which local government was run. This was the city hall, or *casas de cabildo*. Here were the chambers in which most of the *corregidores* lived, and offices in which a variety of magistrates, scribes, and inspectors carried out their functions. On the ground floor of the north side of the building, along the street going down to the Jesuit church, were several shops that the town council leased to tradesmen—the rent providing some of the funds needed to run local government. In 1625 these little shops were occupied by notaries, a baker, a butcher, Indian candle makers and shoemakers, and a general retailer (*pulpero*) called, by chance, Diego Rodríguez de Quiroga. By contrast, on the upper floor of the *casas de cabildo* was the most ceremonial room in Potosí, the chamber of the town council itself. In Antonio López's time, this *sala* was decorated with sixteen portraits of members of the house of Habsburg, a large painting of Christ (with a canopy and a curtain), and three landscapes. The floor was covered with a carpet, on which stood eight benches with quilted upholstery for the comfort of the assembled aldermen (*veinticuatros*). A writing desk, covered with cloth, and bearing a silver inkwell, sandbox, and bell, served the scribe while he took minutes. And two silver maces represented the authority and dignity of the *cabildo*.[65]

To share in that authority was the ambition of many men in the upper crust of local society, since office in a town council was the highest political and administrative position to which many American-born people, unless well connected, could aspire. But possibly more alluring still was the dignity of holding municipal office. To become an alderman (of whom there were only sixteen, despite the title *veinticuatro*) was to have one's social standing displayed and confirmed, and one's financial success recognized—quite literally so since the principal means of gaining these offices in the seventeenth century was to buy them, and some of Potosí's silver producers paid heavily for their *veinticuatrías*.[66] Still greater sums went for one or two other town offices. To be chief constable (*alguacil mayor*) was to exercise great local power, and to command a very large, if under-the-table, income in payments from wrongdoers. In the first decade of the seventeenth century the position fetched 110,000 ducats (nearly 152,000 pesos) at auction.[67] And also especially coveted, though mostly for reasons of ostentation, was the position of royal ensign (*alférez real*), the standard bearer in the many civic acts that were put on regularly in Spanish American towns, and with particular splendor in Potosí.[68] When a new *corregidor* approached the town, half the alderman went to meet him a day's journey away, arranged lodging for him there, provided dinner that night, and lunch the next day before he resumed his progress into the Villa Imperial. The remaining aldermen stayed in town to welcome him with another ceremonial dinner and lunch, and then perhaps a bull fight, a play, and the lighting of bonfires (*luminarias*) around the town. Special bull fights were also offered to important ladies in local society, and to visiting dignitaries—the president or a judge of the high court of La Plata, or a bishop from that city. Sometimes the town council put on a bull fight for its own honor and pleasure. Then again, there might be a procession to intercede with the Virgin for rain, or for the health of the royal family, or of the people of Potosí, or a solemn pageant on the death of a king or queen. In all these, the council, and especially the ensign, had a prominent and public part to play.[69]

If the aldermen of Potosí looked to the *casas de cabildo* with pride and self-satisfaction, they regarded another building, fifty yards away on the lower south side of the plaza, with more ambivalent feelings. Here was the Real Caja, or office of the royal treasury, where refiners had to go to pay the fifth of all silver produced that the crown demanded as royalty in return for allowing them free access to minerals in the earth.[70] And although making these payments gave the miners and refiners a strong sense of being essential to the well-being of the whole Spanish world, a tax rate of 20 percent was more than many in the seventeenth century could bear. The Real Caja, which actually did contain a large chest, or *caja*, for the safe keeping of the king's silver, was the most powerful symbol of royal authority in Potosí; and its three principal officers, the treasurer, accountant, and factor, were, along with the *corregidor*, the senior governmental figures living in the community. Each of the three held a key to the chest, which was made with three separate and different locks, so that all three officials had to be present to open or close it.[71] And convenient to the treasury building—right behind it, in fact, going toward the Rich Hill, though with doors opening onto the plaza—was the Casa de Moneda, or mint, where much of Potosí's silver was struck into coins—coins which, by the early seventeenth century, were found along all the main trade routes of not only the Spanish Empire, but of Europe and Asia too.

On the north side of the plaza, then, stood the Iglesia Mayor, embodying the authority of more than earthly powers; on the south side, the Real Caja and the Casa de Moneda, all too present marks of the authority of the distant king; and on the west side, the Casas de Cabildo, the seat of local power and dignity. By contrast, the upper, east, side of the square lacked any imposing public building or institution. It seems to have been given over to basic, everyday activities of the townspeople. Bordering much of it were shops, some of them facing onto a stone-flagged area called the Empedrado (the rest of the plaza being unpaved, apparently), and others overlooking another lesser open space called the Plazuela de la Fruta. Giving onto the Empedrado in the 1630s was a gaming room, one of many in Potosí. And perhaps in it could have been glimpsed, in among the gamblers, one or two of the hundred and twenty "women in petticoats and veils who publicly occupy themselves in the amorous activity" remarked upon by an observer earlier in the century.[72] But the focus of the upper side of the main square, and the clearest statement of its public purpose, was undoubtedly a large fountain. This six-sided stone basin, just below the Empedrado, received drinking water piped in from springs about a mile away on the eastern edge of the town. Many houses in Potosí had wells from which water could be drawn for washing and cleaning; but the well water was not thought healthy for drinking. So the municipal fountain was a central component of the town's life in more ways than one— a prominent physical object, a crucial contribution to public well being, and, like the plaza as a whole, a natural meeting place for all *potosinos*.[73]

The plaza lay oriented firmly along the four cardinal directions; and from it extended, north, south, east, and west, the main streets of the town, forming a pattern of regular, rectangular blocks. In all this, Potosí conformed to what was laid down by law and custom in sixteenth century Spanish America as the proper urban plan.[74] Somebody (it still remains to be seen who) in the town's earliest years was able to establish this strong gridiron pattern, which persists, even in its details, to this day. By the late sixteenth century, in its physical form Potosí was far from being the jumbled mining camp of huts, furnaces, stables, and refineries, that might have been expected to develop in a place where fortunes could come and go with such speed and unpredictability. Quite the contrary: the town possessed by then thirty or more regular blocks, sixty to seventy yards on a side, interspersed with open squares for market places. The largest of these, larger indeed than the main plaza, was known to Spanish inhabitants as the Gato—the nearest that Spanish could come to the sound of the Quechua word *qhatu*, meaning simply "market." This space, directly northwest of the plaza (fig. 2), was above all a place where silver ores were traded—not the ores that Spanish refinery owners used, but mostly those that Indian mine workers received as wages, or that they extracted, more or less legally, from mines where they were employed.[75] Other squares were usually named after some specific function, such as the Plazuela de los Sastres, or Tailors' Place, one block east of the main plaza; or the Plaza

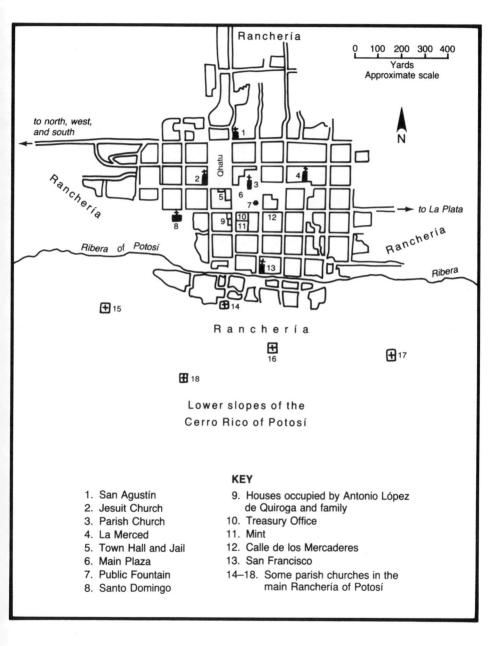

Rancheria

0   100   200   300   400
              Yards
       Approximate scale

to north, west,
and south

N

Qhatu

Rancheria

1

2   3   4

5   6
    7

8   9 10 12
       11

to La Plata

Rancheria

Ribera of Potosí

13

Ribera

15   14

R a n c h e r í a

16   17

18

Lower slopes of the
Cerro Rico of Potosí

KEY

1. San Agustín
2. Jesuit Church
3. Parish Church
4. La Merced
5. Town Hall and Jail
6. Main Plaza
7. Public Fountain
8. Santo Domingo

9. Houses occupied by Antonio López
   de Quiroga and family
10. Treasury Office
11. Mint
12. Calle de los Mercaderes
13. San Francisco
14–18. Some parish churches in the
       main Ranchería of Potosí

A Basic Street Plan of Potosí in the Seventeenth Century
(After the *Planta General de la Villa Ymperial de Potosí*, in *Atlas of Sea Charts*
(K3), Hispanic Society of New York.)

de las Gallinas, where hens, eggs, lard, and charcoal were sold.[76] Very often, squares, streets, and houses would be described by their proximity to some religious building. And indeed these numerous structures, dispersed all over the town, were reference points constantly in the mind of *potosinos*, defining the spaces in which they lived. By the end of the sixteenth century most of the churches, monasteries, and convents that Potosí was to have were in place. Most prominent after the Iglesia Mayor, both on the ground and in the people's minds, were the foundations of the regular clergy: San Francisco, three blocks south of the plaza, bordering the *Ribera*; La Compañía (the Jesuit church), one block west of the plaza, overlooking the Gato; Santo Domingo, three blocks west of the center; La Merced, two blocks east of it; and San Agustín, three blocks north. Thus the establishments of the regular clergy—Franciscans, Jesuits, Dominicans, Mercedarians, and Augustinians—made a ring around the center of the town, where the prime symbol of the secular church, the Iglesia Mayor, stood.

Beyond and surrounding the large and clearly defined area of precisely arranged, square blocks lay, it is true, a series of much less organized suburbs, trailing out into open land. In them dwelt, above all, the native people who provided the town with physical labor. The most prominent, and possibly largest, of these suburbs, or, as they were known, *rancherías*, lay under the slope of the Rich Hill itself, south of the town center, and separated from it by the *Ribera*, the watercourse along which the silver refineries were strung out. Here, indeed, public ways were often no more than jumbled, twisting alleys. And as they went around them, inhabitants of this *ranchería* were always reminded by the heavy thump of trip hammers inside the *ingenios* of the hard labor awaiting many of them in some stage of the long process by which silver was made. But confusion and disorder could perhaps be expected in the *rancherías* around Potosí, because these were places of transition—physical and psychological borders between the Spanish and the native worlds. To them came many people, dragged to Potosí by the labor demands of the *mita*, who expected to be there for only a year. And many indeed left when their forced labor was over; but many also remained, drawn perhaps by the lure of wages to be made in skilled mine work, or in some craft; or simply made unwilling to move again by the effort and cost of a long journey back home.[77] So the *rancherías* were places of adjustment, living spaces where Indians could retreat and recuperate while adapting to the unfamiliar conditions of the Spanish economic world. The suburbs remained, perhaps, more Indian than Spanish in many respects, such as language, dress, foods, and urban form; and were seen as a foreign or different territory by the Spanish dwellers of Potosí, but yet penetrated by Spanish culture and control. The most obvious sign of this penetration was, once again, churches—for as the *rancherías* grew in the sixteenth century, Spanish authorities, and notably the Viceroy don Francisco de Toledo, had imposed on them an organization into fourteen parishes, each with its stone or adobe church. These buildings—San Cristóbal, San Francisco de los Naturales, San Pedro, Santiago, San Benito, San Bernardo, and several more— formed yet another defining ring around Potosí, separating the urban from the rural, and at the same time symbolizing the incorporation of native into Spanish that constantly took place within the *rancherías*, and indeed within the Villa Imperial as a whole.[78]

On the street leading from the southwest corner of the plaza towards the Hill, and the large *ranchería* lying beneath it, Antonio López de Quiroga lived, at least in the later years of his life. This was the Calle de los Contrastes, or Street of the Assayers— an aptly named thoroughfare, since the upper side of its first block was formed by the buildings of the Treasury and the Mint, and assayers had much work to do with the silver being taken to both those establishments for taxing and coining. On the lower side of that block, López possessed in fact two large houses, one in the middle, and the other at the end opening onto the plaza. Symbolically and physically, he could not have been better placed. In Spanish American towns, the more important a citizen was, the closer he liked to live to the town square. And from his houses Antonio López could look out on places full of reminders of his wealth: the Rich Hill, away to the south; the Treasury office, where he had paid millions of pesos in silver royalties; and the Mint, where the products of his mines and refineries had been struck into vast numbers of pieces of eight.

From his corner house, López in old age could also look up eastwards across the plaza toward a part of the town where he had conducted many financial and commercial dealings over the years, and which, indeed, may have been the site of his earliest business in Potosí. This was the Calle de los Mercaderes, where, according to Arzáns, he had first set up shop very soon after coming to the Villa Imperial.[80] By *mercaderes* was meant in this case not mere general merchants, but the men who imported rare and costly items from many parts of the world. Their standing in the community was reflected in the fact that their shops were grouped in the first block off the plaza (in its southeast corner). In 1649, López's first complete year in Potosí, notarial records show that goods received in the town included silk from Calabria, stockings from Toledo, paper from Genoa, crimson baize from Castile, and rough cloth from Quito.[81] These were the sorts of things sold by the merchants in the Calle de los Mercaderes, and possibly the sources of Antonio López's income during his earliest days in Potosí.

TROUBLES AT THE MINT

When Antonio López arrived in Potosí in 1648, however, he must have wondered if there was a promising future in business there after all. For he found himself in a community rocked by scandal—a commotion that rivalled the war of *vicuñas* and *vascongados* a quarter-century earlier in severity, and even exceeded it in some respects, since this new trouble had implications extending far beyond the limits of the town, and even those of Charcas. Although López was able ultimately to find great opportunities in these difficulties, and use them with skill and profit, he must sometimes have seen them, as did most inhabitants of the Villa Imperial, as an unusually dire manifestation of the general decline long apparent in the town's affairs.

The source of the rumpus was fraud in the mint—adulteration

of the silver used to make coin by the addition of excessive amounts of copper. That coin struck in Potosí was low in fineness had been noticed many years before, as early as 1633, in fact, when a royal order to the then viceroy of Peru, the Count of Chinchón, complained that out of ignorance, error, or malice, officials in the colony were falling short of their obligations in this respect. The assayers in the mint at Potosí were duly warned to do better.[82] But admonitions produced no lasting improvement, as would indeed have been expected, since the cause of the fraud lay as much in the inexorably rising cost of extracting and refining silver, and the consequent temptation to adulterate the metal with cheaper additives before it was "marketed" at the mint, as it did in the peculations of individuals. Individuals, though, were a more obvious target for governmental action, and another worried official observed a decade later, in 1644, that the measures taken to uproot dishonesty among personnel at the mint had not been severe enough. Failure to press them home firmly had merely given the offenders greater confidence, conferring on them yet "greater liberty to adulterate the coinage. . . . This town [being] loose in conscience and full of greed, and free and easy in its actions." The only effective course of action to take was to send a completely trustworthy inspector to examine the operation of the mint.[83]

But, asked the viceroy in 1645, as he thought about this recommendation, might not an inspection bring great risks? For the *mercaderes de plata*, or silver traders, who controlled the making of coins, and were presumably among those responsible for the adulteration, were also the largest source of credit for miners and refiners. If the mint were shaken up too severely, and the business of these men destroyed or made less profitable, then great damage might be done to the credit structure of silver production; and that might bring more harm through lowering silver output than gain through the improvement of the coinage.[84] Furthermore, official inspections had always tended to cause trouble in Potosí. "The natural restlessness that the climate of Potosí induces in its inhabitants, and the facility with which they are moved, for very slight reasons, to quarrels and fights which result

in deaths and riots, and commotion in the republic . . . ;" and
again, the hostilities, gossip, rumors, accusations, and so on that
inspections tended to generate; not to mention the "turbulence
and movement, wars and travails" afflicting all the king's realms,
which might incline *potosinos* to indulge more readily than usual in their restiveness: altogether it was more than the viceroy
could comfortably contemplate.[85] Clearly enough, the ghosts of
vascongados and *vicuñas* were rattling their chains loudly in the
halls of his palace in Lima, and apparently also raising lurid visions
in his mind of a Potosí, or whole province of Charcas, trailing Portugal or Catalonia into horrid insurrection.

The public prosecutor (*fiscal*) on the staff of the Council of the
Indies in Madrid, however, was beyond the range of these ghost-
ly warnings, and thought the viceroy's timidity exactly the wrong
approach. Nobody in Spain, he said, wanted to take coin made in
Potosí in business transactions, because its poor quality was notorious; so bad, in fact, that "there is not a *patacón* [piece of eight]
of it that does not contain almost two *reales* of copper." Since a
piece of eight contained eight *reales*, this amounted to an adulteration with base metal of almost 25 percent by weight, from
which "many people in that land have profited by over 500,000
ducats in four or six years." Adulteration of the currency was a
matter of *lèse majesté*, an offence of the gravest sort, which must
not go unpunished. The temperate line suggested by the viceroy
would merely encourage the people of Potosí to commit all sorts
of other crimes. So a minister of great reliability must be sent to
inspect the mint.[86]

The man chosen for the task was don Francisco de Nestares
Marín, a priest, a doctor in civil and canon law, a former Inquisitor in Galicia and Valladolid, and a prickly and puritanical character among whose first acts on taking up office was to issue a
decree forbidding the judges of the high court of La Plata to frequent gaming houses there or in Potosí. Judges of that rank, he
proclaimed, were "great ministers . . . , fathers of their native land,
light of the world, and salt of the earth," who should set a good
example to those around and under them.[87] In reality, Nestares
Marín's jurisdiction was drawn broadly to include far more than

the Potosí mint, for the Council of the Indies had decided that dealing with that problem could usefully be combined with winding up a general inspection of the high court in La Plata, and everything under its legal and territorial control, that had begun in 1633, but which, for reasons of personality conflict, complexity of issues, and excessive scope, had never yet been concluded. And so Nestares was assigned not only the task of restoring the quality of coins struck in Potosí, but also that of resolving large issues (including serious controversy over forced Indian labor) left hanging by two previous visitors. To allow him to proceed with the maximum of authority and expedition, he was appointed to the presidency of the court of La Plata. As such, in the territory of the court, he could be overruled only by the viceroy of Peru, or by an order from Spain.[88]

Nestares spent most of his time as president of La Plata (and most of the rest of his life, for he died in La Plata in April 1660) in Potosí, putting the mint to rights, and attempting to keep the town and its silver industry in good and productive order. He found it no easy or pleasing task. "There is no difficulty, sire," he wrote to the king after being in Potosí three months, "in being a judge in Spain, for there the river knows the rock that guides it . . .; but here rivers are not happy unless they run free."[89] His attempts to set the course of rivers in Potosí indeed brought much friction between him and the citizens of the Villa Imperial. For Arzáns, Nestares's purging of the mint was the third in a series of punishments that God had visited on Potosí for the sins of its inhabitants (the first being the conflict of *vicuñas* and *vascongados*, and the second, the bursting of a large dam above the town in 1626).[90]

Nestares Marín entered Potosí on December 11, 1648. Even the prospect of his arrival had begun to disturb the delicate credit mechanism of silver production in the way that those who had opposed poking into the internal workings of the mint had feared.[91] The town council reported to the king a year later that Nestares had proceeded with zeal and prudence, and that coin was now being struck in accordance with the law. But this had reduced the profits of the men engaged in making coin, who had consequently withdrawn from business—to the detriment of silver producers and commerce in general.[92] This seems to have been a true statement of the situation, and not merely the sort of complaint to be expected from a highly interested body. Nestares had indeed shown much determination, first removing from office the treasurer and the magistrate (*alcalde*) attached to the mint, and then starting an investigation into the affairs of the three silver merchants whom he found most seriously implicated in the fraud. These were don Luis de Vila, Diego Fernán Rodríguez, and, most prominent of all, Captain Francisco Gómez de la Rocha, a man admired in Potosí for his wealth, but also, if Arzáns is to be believed, popular for his generosity.[93] La Rocha's open-handedness had not been restricted, moreover, to good works in the town alone. In 1645, the king, no less, had sent a letter of thanks to him in recognition of his liberality in supplying interest-free loans to the treasury, and in raising at his own cost a hundred and fifty troops for frontier duty at Valdivia in southern Chile.[94] Perhaps la Rocha thought that this sort of service to the crown excused, or compensated for, his manipulations in the mint. A man of Nestares Marín's temperament, on the other hand, can only have seen such behavior as hypocritical double-dealing. And Nestares, the former Inquisitor, was surely further and particularly offended by the fact that this man, the greatest of sinners in the iniquities of the mint, held the office of magistrate of the Holy Office of the Inquisition in Potosí.[95]

As soon as Nestares began to look into the affairs of these three silver traders, and to propose that they should "compose" with the crown by paying large fines in atonement for their alleged criminal fraud, they began to hide their possessions and retired to the Iglesia Mayor. From this refuge they bargained with the inspector, who finally agreed that Gómez de la Rocha should pay half a million pesos over five years, Vila 300,000, and Fernán Rodríguez 200,000. Besides them, Felipe Ramírez, one of the assayers in the mint, was to pay 60,000 pesos; and Nestares was confident of extracting a similar amount from Juan de Figueroa, the senior assayer there. Two others of the accused left Potosí altogether, fleeing to Panama, where they were caught by order of the viceroy, and sent back to Lima. The cash they had taken along in their flight, and goods they had left in Potosí, together worth 240,000 pesos, were confiscated and paid into the royal treasury. Other goods and cash belonging to an alderman of Lima, Alonso Sánchez Salvador, who had once been a silver trader in Potosí, and treasurer of the mint, were seized there, and found to be worth some 280,000 pesos.[96] The total sum at issue in these various confiscations and compositions was therefore around a million and a half pesos: a substantial amount, equivalent to roughly one and a half times the crown's annual royalty income on silver production in Potosí in these years.

In bargaining with la Rocha and his companions in crime, Nestares Marín was following, if unhappily, instructions given to him by the Council of the Indies. The plan was that once punishment had been administered in the form of fines, the same silver traders as before, now chastened and better men, would resume their activities as makers of coin, and, most important of all, as suppliers of credit to the miners and refiners.[97] That overriding fear of upsetting the credit system in Potosí had made the Council unwilling to propose even the minimally severe remedy of excluding these known malefactors from the scene of their wrongdoing—let alone some really radical measure such as putting mintage under direct crown control. In the event, however, Nestares found himself pushed toward more stringent measures than the Council of the Indies had ordered.

Once the size of the fines had been agreed upon, Gómez de la Rocha and his two colleagues emerged from the Iglesia Mayor, and went back to work in the mint. Gómez was even allowed to exercise his inquisitorial office again. (It had been assigned to someone else in March 1649.) Nestares Marín then began to draft rules that would allow the silver brokers enough profit in coining silver to persuade them to stay in the business. At the same time, however, he began demanding that Gómez pay up his first year's fine of 100,000 pesos and provide guarantees for the subsequent payments due of 400,000. At this, Gómez hatched a plot, late in 1649, to murder the visitator by persuading his cook to poison his food with *solimán* (mercuric chloride), or so, at least, the viceroy said

42

CHAPTER I

Nestares had told him in a letter written from Potosí on December 31, 1649.[98] Exactly what happened is hard to decide. On the one hand, it is easy to imagine Gómez de la Rocha being driven by the pressures on him to desperate and irrational measures. Those pressures are hinted at by the fact that even his wife had refused his request, in mid-September 1649, to become a co-guarantor of his debts.[99] On the other hand, it is equally easy to conceive that Nestares, isolated in Potosí among many hostile and locally powerful people, might see plots in the making all around him; and some small carelessness in the preparation of his food might well have made him jump to conclusions about a plot hatched by the man who had most reason to attack him, Gómez de la Rocha. It is perfectly possible also, of course, that Nestares concocted the story of the poisoning so as to damn Gómez once and for all—especially since he had always been dissatisfied with the kid-glove treatment of offenders at the mint ordered by the Council of the Indies.

Elimination of Gómez de la Rocha was, at any event, the course that Nestares Marín now followed. Some time in January of 1650 Gómez was executed by garrotting in Potosí, and his corpse hanged in the plaza "as a public example." Nestares now proceeded, indeed, to the severe and exemplary measures that had always been his own preference. He also executed Felipe Ramírez, one the assayers at the mint, who, he now decided, was the greatest offender in that institution. And he arrested don Luis de Vila, one of the silver traders who had taken refuge along with Gómez de la Rocha in the Iglesia Mayor, and who had subsequently emerged once the size of the fines had been negotiated. Nestares now decided that Vila's guilt was sufficient to warrant his arrest; and when Vila's supporters removed him from jail to the supposed sanctuary of the Augustinian monastery in Potosí, Nestares simply plucked him out again and put him back in prison.[100]

The final outcome of Nestares's rigor was that by mid-1652 only three silver traders were still coining silver at the mint in Potosí, of a dozen or so who had been in business before the inspection.[101] The fines imposed by the visitator, combined with his demands for higher-quality coinage, had driven the others from the business. The effects of the inspection, if not as dire as the prophets of doom had foretold, and not the third great disaster proclaimed by Arzáns, were still considerable. Silver production dropped sharply from 1649 to 1651, though it stabilized after then for several years. (See Fig. 1.) The inhabitants of Potosí suffered from a reduction of the circulating value of coins struck in Potosí that was put into effect in May 1652—a reduction ordered not so much because of anything that Nestares had found as because of the well-known low fineness of Potosí coinage that had prompted his inspection in the first place. Pieces of eight struck in Potosí before the start of the inspection in 1649 were to circulate temporarily at a value of 6 *reales* only, or 75 percent of their face value; and those struck after that date, in Nestares Marín's time, were to circulate at 7 1/2 reales, or 93.75 percent of face value. By the end of October 1652 all these coins were to be remade, at legal fineness, in the mint, and after that date, none of the old coin should have any monetary value.[102] Since most coin held by *potosinos* would obviously have been made before 1649, many stood to lose severely by this devaluation. Those losses are perhaps partly to blame for the grim account of Potosí given by the *corregidor* a year or so later. The condition of the town was, he said, "the most miserable that has ever been known, with bankruptcies of merchants [here] to the tune of more than a million and a half [pesos], and consequent very great harm to those of Lima and Spain." Producers of silver had been pressured by Nestares Marín to pay off old debts that they owed to the treasury, for such things as sales taxes, and surety bonds they had issued long ago, and to do so they had been forced to accelerate processing of their ores, which had meant inefficiency and losses. To raise yet more cash to meet their debts they had resorted to selling off their belongings. And in many cases the treasury had simply taken over the refining mills, and leased them out, using the rent as payment on the debts. And it was well known, the *corregidor* noted, that refineries were irreparably damaged by forced leasing, and that "the silver producer who falls never raises himself again."[103] Nestares Marín himself acknowledged that things had become difficult in Potosí when he wrote to the king, at about the same time, that the devaluation of the coinage had led to a general "unccrtainty and confusion of spirits" in the Villa Imperial.[104]

But the greater the confusion, the greater the potential gains for those who could keep a cool head amidst it. One such was Antonio López de Quiroga. The troubles undoubtedly had damaging effects on any general trading that he did during these years. But much else was to his advantage. As a man recently arrived in Potosí, he had no old debts to pay off to the Treasury. He had no personal memories of an easier and better past in Charcas to embitter or discourage him. He had not owned mines or refineries that he was now forced to dispose of in a buyer's market. He was, indeed, himself far more in the position of a buyer in that market than of a seller. He had married into the family of a most substantial and repsected merchant and moneylender in the community, Lorenzo de Bóveda—a man, moreover, apparently untainted by the scandals surrounding the mint. And although it is true that some of Bóveda's wealth must have been held in coin subject to devaluation, and that therefore whatever had come to López by way of dowry or inheritance would have suffered some loss in value, nonetheless the marriage can only have conferred on him considerable profit, and notable respectability within a very few years of his move to Potosí. All this stood to benefit him, if put to wise and productive use. And indeed he did put it to such use, launching a career in the early 1650s that rarely, if ever, faltered thereafter.

2

ESTABLISHMENT

MERCADER DE PLATA

THE CRUCIAL STEP THAT ANTONIO LÓPEZ TOOK ONCE HE WAS SET-
tled in Potosí—the step that set him on the road to great wealth
and unrivalled dominance of silver production—was to try his
hand as a *mercader de plata*, or silver trader. Since this was the
business that had proved the death of Gómez de la Rocha and
the ruin of many others in the 1640s, it was a risky move to make.
And indeed for Antonio López it was to prove a costly venture—
in cash, at any rate—if his own account of his fortunes in the
1650s is to be believed. But his experience as a *mercader de plata*
brought him into contact with the processes of silver making in
Potosí (something about which he apparently had no previous
knowledge), and especially seems likely to have provided him with
an education in the capital structure and operations of the industry. Furthermore, although by the late fifties, after several years
of experience in the silver broker's trade, he complained of being
burdened with many bad debts, he had begun to emerge as a powerful figure on the financial scene of Potosí, precisely because he
was owed so much money. In other words, then, his activities as
a silver broker, if expensive when measured in cash (though quite
possibly he does not mention profits offsetting those bad debts),
carried him to a position from which he could advance far more
easily into silver production itself than if he had tried to do so
immediately after arriving in the Villa Imperial.

In Spanish American silver mining, two large functions gave

the *mercaderes de plata* a position of central importance. The first was to organize the conversion of raw silver into coin, so making the metal usable as money. To do this, they bought bulk silver, as it emerged from the refining process, from the owners of ore processing mills. They then took it to the Treasury office, where they paid the royalty tax of a fifth along with other minor taxes. And from there they transferred the silver to the mint, where the quality of the metal was adjusted, in furnaces owned by specialist craftsmen, to the standard required by law for coinage, and formed into thin strips (*rieles*) from which coins could be struck. The blanks were cut, and the money stamped, by coin makers (*acuñadores*) appointed by the treasury.[1] The particular capacities provided by the *mercader de plata* in this process of making coin were, first, the ability to put up large sums in cash to buy raw silver from the refiners; second, knowledge of the many bureaucratic requirements that had to be fulfilled during mintage; and third, enough technical expertise to oversee the metallurgical work of the furnace operators and coin makers, all of whom had good opportunities to cheat both the crown and the *mercader* at various stages of the coining process.

The other function of the silver traders was to provide credit for miners and refiners. (Credit in the mining context was known as *avío*, and anyone who gave it was known as an *aviador*.) *Avío* might take the form of cash loans, or of deferred payment on the supply of raw materials, tools, and other goods needed to make silver. Credit was given, of course, at interest. The difficulty is to know what the rate of interest was. Loan notes in Potosí in the seventeenth century often contain the phrase *"sin interés alguno"* ("without any interest at all"), which was clearly not so, and possibly inserted to meet legal and ecclesiastical objections to excessively high interest charges. The normal practice seems to have been to deduct some proportion of the face value of the note at the time it was granted, and then to require payment at full face value when the date for repayment arrived.

It is, in fact, very hard to estimate what the rates of profit of the silver traders amounted to, from either their minting or their lending activities. What can be said is that these men usually seem to have secured for themselves (not only in Potosí, but generally throughout the silver producing regions of Spanish America) a large share of the gains available from mining and refining silver. Perhaps this was partly because they had, as the saga of the Potosí mint in the 1640s shows, excellent opportunities for profitable fraud, for which they were prosecuted only when they carried it to extremes, as Gómez de la Rocha and his companions had done. Perhaps it was because they could afford to be judicious and selective in granting loans, whereas miners were in an occupation with intrinsically greater financial risks. But whatever the cause, in Potosí as in other mining towns of the Andes and Mexico, the *mercaderes de plata* tended to accumulate more wealth than the men directly occupied in mining silver ore and turning it into metal. They were not immune, naturally, to general alterations in the prosperity of mining. *Mercaderes de plata* in Potosí undoubtedly had an easier time of it in the late sixteenth century than in the mid-seventeenth. But, with caution and effort, some of them, at least, still seem to have been able to turn an honest profit even amidst a general situation of falling silver output.

López himself provides two different, though not incompatible, accounts of how he became a *mercader de plata*. According to one of these, he used the considerable capital and good name he had quickly gathered in Potosí to set up another man, Captain Juan de Orbea, "a person of complete understanding of the business of silver trading, making coin, and financing the refiners and miners of Potosí and its surroundings," as a silver trader in December 1650. Orbea ran the business until his death in 1655, at which point López "carried it on himself, assisted by his stewards and servants, with no small labor on his part, and with great profit to the service of his majesty."[2] Juan de Orbea is indeed referred to by the *corregidor* of Potosí as one of the three silver traders at work in 1652,[3] while Antonio López is not identified as a *mercader de plata* until five years later. So there is some independent confirmation for at least part of this story, although who Orbea was and why López should have collaborated with him as he did, are questions on which no evidence has emerged.

While in this first account López presents himself as an independent actor, his second, and slightly later, explanation of how he became a silver trader assigns a role also to that looming figure of mid-century Potosí, Dr. don Francisco de Nestares Marín. Here the possibility of yet another Galician connection must at least be raised, because for several years before 1640, when he was appointed an Inquisitor in Valladolid, Nestares had served the Holy Office in Galicia, first as a prosecutor, and then, after 1637, as Inquisitor. He had been an ambitious man, always anxious to establish profitable connections,[4] and it is likely enough that in his Galician years he had crossed paths with some of López's relatives.

López, to be sure, never suggests any earlier contact with Nestares Marín, either direct or indirect. He merely notes in his second account that he and the visitator arrived in Potosí at the same time (late in 1648). His story then continues with an unexceptionable recital of Nestares's reforms in the mint, and remarks on the consequent shortage of people willing to undertake manufacture of coin. At that point, says López, Nestares, recognizing the difficulties that his severity had created (restricted credit for mining, reduced extraction of ore, and a fall in royalties),

proposed to me several times the pleasant service that I would render to Your Majesty by minting silver. At that time I found myself with some wealth (and the greatest part of it was the truthfulness and plain dealing that I have always practised, on account of which I have had charge of other people's fortunes), which was the subsidiary reason to the principal one of serving Your Majesty that encouraged me to embark on a business of such importance. I began to cast silver for coinage in the year of 1652, from which fact the said President took much pleasure, saying publicly that it was the greatest service performed for the king in this realm [of Peru]. . . .[5]

What these accounts, and the discrepancies between them, would seem to add up to is that López gradually eased himself into business as a *mercader de plata* in the early 1650s. Quite possibly Nestares Marín did encourage him to go into silver trading, and on that prompting he initially experimented with it as the silent partner of a man, Orbea, already knowledgeable in the trade. Quite possibly, also, however, López needed only a slight push to induce him to try his hand. The opportunity was certainly there for anyone with capital and commercial acumen—both of which he seems to have possessed in some measure by the opening years of the 1650s. With most of the previous *mercaderes* swept away by Nestares's broom, shortage of loan funds may have raised the interest rate that *aviadores* could demand from miners and refiners. And Nestares himself saw the need to improve the return that silver traders received in organizing mintage. To that end he slightly reduced the lesser levies that were charged on silver presented for taxation and coinage, thereby increasing the amount of raw metal remaining in the traders' hands.[6]

The idea that Antonio López gradually eased himself into the silver brokering business is supported also by the Treasury and mint records of Potosí. They show that he began to deliver bars of silver for taxation at the Treasury office in April 1653. Since there is absolutely no evidence that he was producing silver from any mines or mills of his own at that early date, these bars were most likely silver that he had received from refiners as repayment of principal and interest on loans that he had made to them. It was the normal practice for the creditor to take repayment in untaxed ingots, which he would then carry to the Treasury for deduction of the royal fifth. So López, by early 1653, may well have been at work as an *aviador*. This was the less complicated and technical of the two functions of the *mercader de plata*; and López already had years of experience in lending money. In contrast, the second function, supervising minting of coins, demanded considerable knowledge of specialized law and of the technicalities of fineness and assaying. López did not, according to the mint accounts, launch into this until two years later, in April 1655, shortly before the death of his partner, Juan de Orbea, which happened in June of that year. It was in that April that he first began delivering ingots of silver to the mint for melting down into the strips from which coins were struck.[7] Now he was a complete *mercader de plata,* and by the end of 1657 the viceroy was referring to him by that title.[8]

Throughout these years of learning to be a silver trader, López also worked at more general commerce. By early 1654 he was consistently identified as a *mercader,* which strongly suggests that he had become one of Potosí's importers. In October of 1654, over four days, he despatched to Lima a total of 29,000 pesos for delivery to correspondents there who were to dispose of this very considerable amount according to his instructions.[9] No record remains, alas, of what these instructions were, but an identified *mercader* sending sums of that magnitude to the terminus of the import trade from Spain (and the Orient) is very likely to have been buying goods arriving from overseas. At the same time, López was active as a business agent for other people. In September of 1654, for instance, he was empowered by another *mercader* of Potosí, Gaspar Liano de la Vega, to collect 3,150 pesos that Liano had lent to one Toribio de Vargas in the mining camp of San Antonio del Nuevo Mundo, far off in the south of the Potosí district. Vargas had evidently left San Antonio for Cuzco by this time, and particularly interesting in this transaction is the fact that López called on two agents of his own in that city to collect the debt—an action that suggests that he was on the way to establishing his own network of business correspondents in the trading centers of the central Andes. Even more suggestive is that one of his representatives in Cuzco was the *alguacil mayor,* or chief constable of the city— a powerful figure in any municipality— and, to judge by his name (don Antonio de Losada y Novoa) possibly a relative of López's.[10] In that same year of 1654 and the next, López handled money passing through Potosí on the account of at least three other people in and around the Villa Imperial. One of them was don Juan de Somoza Losada y Quiroga, previously the governor of Santa Cruz de la Sierra, and now living in a mining camp called Espíritu Santo, northwest of Potosí. This was the same Somoza Losada y Quiroga that Antonio Lopez had represented in Lima before the viceroy in January 1642.[11] While collecting and despatching sums belonging to other people was far from being unusual business for an established merchant to undertake, López's activity in this line is evidence that he was acquiring a reputation for that "truthfulness and plain dealing" of which he was later to boast.

A further sign of his rise in public regard in his early years in Potosí was his election, late in 1652, as *síndico*, or temporal agent, of the Franciscan monastery there.[12] This was the oldest monastic foundation in Potosí, dating back to the 1550s, and a prominent one in the minds of the citizenry. The position had been held by Antonio López's father-in-law, Lorenzo de Bóveda, up to his death in mid-1652. Undoubtedly that connection weighed with the Franciscans as they cast around for a successor to Bóveda. But, notwithstanding the family link, to be appointed syndic was to receive a mark of recognition from a highly respected and influential local institution, and can only have bolstered Antonio López's budding reputation. The post was, moreover, more than merely honorary. The syndic managed the affairs of the monastery in the outside world, administering its properties, receiving fees for such services as burials and the saying of masses, collecting debts, alms, and rents, providing legal representation, and maintaining the monastic buildings. In June of 1654, for example, López supervised the purchase of 2,000 pesos' worth of timber for the roof of the principal chapel.[13] Although in 1670 the press of his own affairs made him give up handling the Franciscans' business personally,[14] he was to remain *síndico* until his death twenty-nine years later. More than that, indeed, late in life he became syndic of all Franciscan houses in the province of Charcas.[15] And it was among the Franciscans that he was buried, in a vault that he had built for himself and his heirs beneath the Altar of the Immaculate Conception in the conventual church in Potosí,[16] still the servant of the foundation that had publicly bestowed trust on him so early in his career in the Villa Imperial.

The Antonio López de Quiroga of the early 1650s, then, was busy with many things besides learning to be a *mercader de plata*, and they were things that he pursued in greater or lesser measure throughout the rest of his life. But silver trading was the track that would lead him to his distinctive wealth and power in Potosí. He can hardly, though, have had any such hopes at the time, at

least if his own telling of the fortunes of his initial business partnership with Juan de Orbea is to be taken at face value. Looking
back many years later on those early days, he recalled great difficulties: when Orbea died, in mid-1655, he had lost, in unrecov-
erable *avío*, 150,000 pesos' worth of López's money. This loss,
López noted, was partly the result of droughts in Potosí in 1651
and 1653 that hampered milling of ores, and hence limited the
production of silver and the ability of the refiners to pay their
debts. Perhaps partly because these problems had natural origins,
which were not likely to recur soon, López decided to press ahead
with the *avío* business after Orbea died. He also, by his own statement, thought that he could meet financial obligations that he
himself had incurred only by carrying on as an *aviador*. But further troubles awaited him, with the result that by the end of the
1650s (so it would seem) he said he was faced with a bad debt of no
less than 870,000 pesos.[17]

Now it is true that in the early 1650s, Potosí was afflicted by a
drought that depressed silver output. But the rest of López's protestations have a ring of hollowness to them—indeed, a booming
resonance more than a ring, since the statements that surround
his assertions of great losses are so clearly designed to show him
as the proud and self-sacrificing hero of Potosí in that decade. One
means he uses to achieve that effect is to belittle Juan de Orbea.

[He], with my silver and the coin I supplied to him for that purpose,
granted the said *avíos* and made very substantial payments into the Treasury on behalf of the miners and refiners, for mercury [they had bought]
and other debts owing to his Majesty. . . . And if the Indian [workers]
and mines had not been sustained by [the funds] supplied at that time
by the *aviadores* and in particular the said Juan de Orbea, it is very certain that the *mita* would have collapsed. And all this was attended to
for the most part by the said Juan de Orbea, using the money that I gave
and supplied to him. And he remained in this business until June 4,
1655 when he died [owing López 150,000 pesos and leaving him no choice
but to go on with the business to try to recover them], . . . which I did
from the day of Juan de Orbea's death. And so it is fully evident and
certain that I was the *aviador* and payer of Treasury debts from the said
year of 1651 for the entire refining industry of Potosí and the surround-

ing mining centers, because the said Captain Juan de Orbea was no more than a person put in place by me and who traded with my money and wealth, and his fortune was very slender and slight when he entered on this business; and so all the said *avíos* and payments to the Treasury are on my account, and are meritorious services of mine alone, performed for his Majesty and the common good. . . .[18]

So Orbea was, it would seem, little better than an incompetent underling, who did no more than take López's magnanimously bestowed funds and squander them. López conveniently fails to recall that he himself, according to Treasury records, had begun to participate directly in the *avío* business in 1653, and therefore bore some direct responsibility for any losses occurring before 1655. He also neglects to bring up in this passage the far larger loss that he claims to have suffered after Orbea died—when his bad debts grew from 150,000 to 870,000 pesos. To have done *that* would certainly have made Orbea look better, and López less heroic; though in another context he could present that greater loss as yet more striking evidence of valiant self-sacrifice, not in suffering and overcoming the effects of someone else's incapacity, but in braving grave and general adverse circumstances.[19] If Antonio López lost his own money, he had done noble service to crown and community; if someone else lost it for him, he was noble in his toleration and perseverance in the face of such ineptitude.

Furthermore, whatever López's losses may have been as a silver broker and creditor in the 1650s, his career and fortunes in general showed little sign of faltering. By April of 1659 he had gained the honorary military title of *Capitán*, which can only have been an indication of rising social standing.[20] Two years later he bought a house, or houses, near the center of Potosí, on the street between the plaza and San Francisco.[21] Both these acquisitions, though, were outweighed by his earlier purchase of what was, by local standards, a large and costly rural estate outside Potosí. This consisted of several pieces of land, or *haciendas*, called San Pedro Mártir, in the Cinti valley, eighty-five miles south southeast of Potosí. López paid the weighty sum of 52,000 pesos for San Pedro, and the twenty-six black slaves on it in March 1658.[22] He kept the estate for the rest of his life, adding land and further slaves to it, and producing wine, brandy, cattle, and food crops there. This purchase can hardly have been the work of a man critically in debt, as López would seem to try to say he was in those years. Even if he did not pay cash for San Pedro, his financial standing in Potosí must have been quite solid if he could raise a loan of 52,000 pesos.

In his own account of his activities during the 1650s, of course, López, to present himself in the best light, stresses his losses while making no reference to his gains, and there is no means of knowing for sure what those might have been. But again the records of the Treasury in Potosí provide some illuminating hints. Between December 20, 1650, and August 22, 1659, Orbea and López (López alone, of course, after June 1655) presented 13,893 bars of silver, worth 14,186,739 pesos and 6 reales, for taxation at the Treasury in Potosí. [23] This, it seems clear, was silver that they had received from miners and refiners to whom they had granted *avío*, and represented repayment to them of the principal they had lent, along with whatever interest they took on *avío* loans. Now that interest rate is not known, but interest of merely 5 percent contained within that total amount presented for taxation would amount to some 675,000 pesos—a substantial proportion of the 870,000 pesos of bad debts that López complained of incurring during the decade; a rate fractionally above 6.5 percent would have yielded enough gain to cancel the bad debts completely. Such an interest rate is well within the bounds of possibility, and a higher one would not be surprising. So it is entirely possible that even while sustaining very large losses in the form of bad debts, Antonio López more than offset them with interest collected on the enormous sums that he lent to producers of silver in Potosí during his first decade in the Villa Imperial. The Treasury records leave no doubt of the scale of his credit operations. Those 13,893 bars of silver that he and Orbea delivered to the Treasury between December 1650 and August 1659 amounted to about 41 percent of all silver brought to the Potosí Treasury for taxation between the beginning of 1651 and the end of 1659.[24] López exaggerated in claiming that he had been the sole financier of mining in Potosí and its environs during the fifties, but he was certainly a bigger financier than anyone else.

MINER AND REFINER IN POTOSÍ

On March 25, 1657, Antonio López de Quiroga took a three year lease on a royal mine in a vein of silver ore called Nuestra Señora de Guadalupe. The mine was at a place called Tabacoñuño, a few miles east of the Rich Hill of Potosí, where silver had been extracted on a modest scale since the sixteenth century. The total rent was a mere 130 pesos. This was López's first move toward producing silver.[25] A month later he rented another royal mine, this one perhaps in the Hill itself, again for three years, for a total of 100 pesos.[26] There is no evidence of his taking any more mines until 1661.

Why did Antonio López venture into mining? The only explanation that he offers, written retrospectively in 1674, has a certain plausibility at first sight if his protestations of great debts and poverty by the late fifties are to be believed (and they probably should not). He presents his entry into mining as a matter of desperate remedies, of dire last resort:

And seeing myself afflicted and unable to pay what I owed, I had to apply myself to working mines in the Hill, excavating a place in it called Amoladeras, where ores of quality had never been found. Everyone mocked this effort as futile, but God was pleased to allow me to strike veins by means of an adit, which gave me rich ore with which to pay my debts. . . . [27]

For various reasons, both specific and general, this will not quite do. To begin with, his first mine was not at the Amoladeras site in the Hill of Potosí, but at Tabacoñuño. This is a small point, however, and he may later have forgotten about that first lease, which certainly never features again in his career as a miner. Second, he may have been mocked for devoting himself to the Amoladeras site on the Rich Hill—but only by the ignorant and foolish. For this place, half way down the west side of the Hill, while not one of the sites of fabulous riches in Potosí's past, certainly had a reputation for high potential. In 1620 it had in fact been called a "place of great wealth," and a scheme had been touted for driving an adit into the Hill precisely there to exploit that wealth.[28] More recently, relatives of people known to López had taken good ores from the Amoladeras.[29] So, in all probability, it was disingenuous of him to imply that he had approached the Amoladeras mines amidst wholesale predictions of failure.

More generally, even if Antonio López's circumstances had been as straitened as he announced, why would he have chosen mining as the solution to his problems? The cause of his losses, after all, was precisely the insolvency of silver miners to whom he had made loans. If they, most of them experienced in producing silver, could not make money at it, how could he, a man totally lacking in practical, personal experience of that sort, expect to do better? He was a man evidently alert to take advantage of opportunities; but he had given no signs to this point of being foolhardy, and it would have been just that for him to expand into mining in the way and circumstances in which he said he had done so.

His tale of success with the Amoladeras deposits, then, while having some relation to reality, as will become apparent, is best regarded as another product of his braggadocio: it pleased him to present himself as the hero who had overcome monstrous losses as a silver broker by plunging boldy against all advice and odds into an unpromising mining venture that, with God's help, concluded in triumph.

The documentary record suggests, though, a far more measured story, comparable with his gradual entry into silver brokering. After his lease of those two mines in 1657, there is no further sign of his interest in making silver for two years. Then, in April 1659, he contracts to purchase two axles for ore mills. These great shafts, 21 feet long, and each costing 1,000 pesos, were to be brought from Jujuy, hundreds of miles to the south.[30] One of them, at least, may have been intended for a broken down silver refinery that he seems to have bought cheaply at about this time at Cantumarca, a couple of miles below Potosí on the *Ribera*.[31] The other may possibly have been destined for another refinery within Potosí, working but in worn condition, over which he gained a large degree of control in December 1658. The owner of this *ingenio*, Bartolomé de Uceda, had borrowed some 10,000 pesos from López in the years up to 1655; by the time of his death, in 1657, the debt had tripled. López then agreed with the executor of the estate that a manager should be placed in the mill, who should continue to operate it, using the silver produced to pay off what López was owed.

This was a typical silver refinery of seventeenth century Potosí, similar to the many others that López was later to own and run (see pl.1). It was on the *Ribera*, in the parish of San Sebastián, a few hundred yards southeast of the plaza. Within the rectangular outer wall, built of adobe, the most striking pieces of apparatus were two stamp mills (*cabezas de ingenio*) for crushing ore, one on each side of a large water wheel, and driven by a single shaft passing through the center of the wheel. Each of these mills had seven stamps—again, a standard number in Potosí—and each stamp was fitted with an iron shoe weighing some fifty pounds. For the actual refining of the ore, once it had been finely crushed under the stamps, were eighteen rectangular tanks, or *cajones*, set in the ground and lined with stone, and each capable of taking 5,000 pounds of ore, together with the reagents (water, salt, and, most important, mercury) that were needed for the silver to be extracted. Some eighty tons of silver chloride and silver sulphide ores lay in heaps within the *ingenio*, waiting to be refined. Various spare, or worn, components—a wooden gear wheel, iron straps of different sizes for reinforcing the wooden parts of the mill, stamp shafts and shoes, paddles for stirring the ore-mixture in the tanks—were stored in rooms built up against the surrounding wall of the refinery. And attached to the wall on the outside were Uceda's living quarters (*casas de vivienda*). This was a slightly unusual arrangement. Normally the owner lived in apartments inside the wall that faced onto the interior work space of the mill. Often a chapel was added to the house, although there is no mention of one in this case.[32]

With occasional interruptions, this mill seems to have continued working under López's effective control for six years. He,

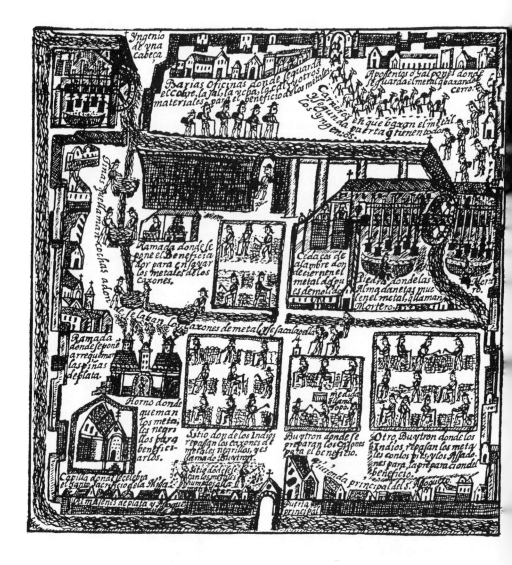

Plan of a silver refining mill in Potosí, c. 1700

apparently, would have liked to continue operating it with an administrator indefinitely. But in 1664 the *corregidor* of Potosí, acting in his capacity as judge in mining disputes, ordered the whole of Uceda's estate to be sold, to meet outstanding debts. López received almost 50,000 pesos, or nearly the entire proceeds, from that sale.[33] It would certainly seem, then, that he got back what Uceda had owed him in 1657.

The fact, however, that he would apparently have preferred to go on operating the refinery himself suggests that the arrangement suited him well. And this may indicate something about his movement into silver production. By 1660 he seems to have had two refineries at his disposal: one that he had bought at Cantumarca, and the other that he operated through an administrator in Potosí. Neither of these had cost him much—at least, not in new outlays. The 7,000 pesos or so that he had paid for the plant at Cantumarca[34] was a small fraction of what he had laid out on the San Pedro estate, and he was able to make use of Uceda's mill in Potosí without further cost (although, of course, he had already in effect invested heavily in the plant through the *avío* he had given to Uceda). In sum, then, López may well have been able to begin a refining career at a low initial cost and risk. The indications are, therefore, that rather than launching himself boldly into a risky adventure in the Amoladeras mines of the Hill of Potosí, he felt his way cautiously into silver production by leasing, in 1657, two royal mines, and processing the ore raised from them in mills that cost him little in extra investment. Once having tried the water, and found it welcoming, he *then* developed his mining activities energetically, and did indeed turn to the Amoladeras site with great determination early in the 1660s.

By the time he began to work las Amoladeras, López had risen to general recognition as a silver producer in Potosí. It was in mid-1661 that he was first called an *azoguero*, or "mercury user" —the descriptive title given to men who produced silver by amalgamation in the Villa Imperial, and usually indicating someone who had a substantial interest in both refining and mining.[35] López had certainly hit a steady stride by that time. The Treasury records for 1661 show that he bought 13,000 pounds of quicksilver from the crown in that year, enough to produce at the mercury loss rates then prevailing in Potosí some 7,800 pounds of silver (or 4.5 percent of the total registered output of the Potosí district in that year).[36] The amount of mercury that he received in 1661 was close to a twentieth of the total coming into Potosí that year and, since the number of *azogueros* active in those years was about sixty, the indications are that López was already more active in producing silver than most of his fellow refiners.[37] And indeed as the decade wore on, the proportion of the mercury entering Potosí that went to him continued to rise—a trend reflecting his growing dominance of silver production.

Just as López seems to have acquired refining plant cheaply and cautiously, so he gathered mines. There is no indication that, at this or any later date, he invested much time or money in prospecting. His method was to rework claims in known ore deposits. Abandoned workings might legally be claimed by anyone interested in working them, and it was by this method that López acquired an interest in the Amoladeras deposits. On August 1, 1661 he went to the mining magistrate, or *alcalde mayor de minas*, in Potosí to put in a claim for his first mine there. The working that he wanted lay deep within the Hill, branching off an access gallery, or *socavón*, more than 200 yards in from the surface of the ground. Having noted the claim, the *alcalde mayor*, in accordance with the law, had it proclaimed publicly three times, on that same day, and on the 5th and 9th of August, so that anyone wishing to object, or assert his own title to the working, could do so. After the third proclamation, the *alcalde mayor* called on Antonio López to produce witnesses to confirm that the working was vacant. He brought four. Two said that the mine had not been worked for more than two years, and two that it had been abandoned for over eighteen months. And finally, on September 1, 1661, the working, sixty *varas* square (about fifty-five yards) in the horizontal plane, and extending indefinitely in the vertical, was adjudicated to Antonio López.[38]

He built rapidly on this small beginning in the Amoladeras. Since an individual could hold only one claim in any single vein, López now registered adjacent workings in the names of relatives and others. Where a different vein could be discerned, however, he could register a mine in his own name. So in January 1662 he took three mines in veins in the Amoladeras named San Andrés, San Antonio Abad, and Nuestra Señora de la Limpia Concepción. Early in 1664 he took a sixty *vara* claim in the Amoladeras for his young daughter, doña Lorenza de Quiroga, and another for one of his sisters-in-law, doña María de Bóveda, whose affairs he administered.[39] As he encountered, or at least identified, new veins, besides registering his own mine in each of them he also, as the law demanded, set aside an adjacent claim for the crown. These deposits he then immediately leased back from the Treasury. In 1667 he was working at least six such royal claims in the Amoladeras. With rents of only 33 pesos a year for each working, the cost of acquiring these claims was trivial.[40]

By the late 1660s, then, López had assembled a large number of claims in the Amoladeras section of the Hill of Potosí—certainly well over a dozen of them, and possibly substantially more. Most, if not all, of these were ore deposits accessible from pre-existing workings. This being so, what was the reason for López's unprecedented success with them? The answer seems to lie in a new orderliness or rationality of exploitation. López's originality seems to have been in combining these neighboring workings into a unit, and then to have exploited the unit through numerous capacious horizontal galleries (technically called adits, or *socavones* in Spanish) that facilitated ventilation, drainage, and access to the bodies of ore.

Adits entering the Amoladeras part of the Hill already existed at the time López began his operations there in 1661. Indeed, the first working that he claimed was one to which there was no access except along an adit. But he did not own that gallery. Rather, the first of these access tunnels that belonged to him seems to have come into his hands by chance. In January 1662, one doña Paula de Figueroa, a widow of good social standing in Potosí, declared that she had received from Antonio López "many very good works and much aid for my sustenance and preservation, and relief and remedy in my needs," in gratitude for which she donated to him all the mines and mining works, rights and shares that belonged to her as heir of her father, don Diego Rodríguez de Figueroa, some-time alderman of Potosí.[41] Among the workings that López so fortuitously received was an adit, called "Mala Moneda" ("Bad Coin") in memory of Rodríguez de Figueroa, who for some unexplained reason had been given that unflattering nickname. He had started the adit in 1630,[42] but did not work it for very long. This old gallery now became the starting point for López's rationalization of underground works in the Amoladeras. He enlarged it during the 1660s, and interlinked it with several other galleries giving access to his ore workings. Exactly how these were laid out is impossible to say. Indeed tremendous confusion reigned at the time over the names and location of the galleries, and gave rise to fierce lawsuits. But an examination of López's subterranean operation at las Amoladeras by the official inspectors of mines in 1676 revealed that he possessed one adit called the "Socavón Grande" that penetrated 550 yards into the Hill, another called "de Amoladeras," a third called "de Cabezas," a fourth called "Chipta," a fifth called "Copacabana," and one more without name. These six galleries were, it seems, all of the statutory dimensions (roughly seven feet high by six feet wide), and interconnected beneath the ground. Many years later, the president of the high court of La Plata visited the "Amoladeras" adit, and found it "remarkable," in part because it had brought López wealth with which he could then launch out into other projects, and in part because it had so effectively drained and ventilated all the veins of silver ore that it had cut. Lack of air was as serious a difficulty as water, because without plenty of it, workers could not function and torches would not burn. Adits also brought the advantage of simple, walking access to mines, in place of a risky and laborious descent of 500 or 1,000 feet down a series of swaying ladders made of round wooden rungs inserted into woven leather cables.[43] In noting all these gains, the president put his finger quite accurately on many of the benefits that adits could bring to mines and those who worked them.

Through his six adits López extracted great hauls of rich silver minerals from his many mines in the Amoladeras, which seem to have been driven into a sheet (manta) rather than a vein of ore.[44] The prevailing type of mineral that he found seems from passing references to have been *metales pacos*, silver chlorides colored a reddish brown by the iron oxide in them, which generally gave fewer problems to the refiner than the sulphides (*metales negrillos*) that also lay in abundance within the Rich Hill. By Arzáns's account these *pacos* provided yields the like of which had not been seen in Potosí since the early decades of its existence: 128 ounces of silver from a *quintal*, or 100 pounds, of ore.[45] The common yield in mid-seventeenth century Potosí was perhaps 1 to 2 ounces per *quintal*. In 1675 the husband of one of the descendants of Diego Rodríguez de Figueroa, who wanted to rescind the gift of the "Mala Moneda" adit made to López in 1662, asserted that he had taken out through that gallery ore yielding no less than ten million pesos.[46] Though this figure may well be an aggrieved exaggeration borne of jealousy or regret (the plaintiff noted how "rich and esteemed" López now was), it reinforces the impression given by other sources that López did extremely well from las Amoladeras.[47]

The best available measure of the scale of his silver production during the 1660s is the amount of mercury he bought at the Treasury during that decade. The total from the beginning of 1661 to the end of 1669 amounted to 308,300 pounds, enough to produce about 200,000 pounds of silver, which were worth, after payment of royalties, slightly more than 2,500,000 pesos. More revealing than these absolute figures, however, are the proportions of the total mercury available in Potosí that went to López. In the course of those nine years, 2,314,200 pounds of quicksilver entered the Treasury in Potosí for distribution among the *azogueros*. López's 308,000 pounds constituted 13.3 percent of this gross amount. Since nearly all (if not all) silver was produced in Potosí by processing with mercury, it would seem, other things being equal, that López's share of silver production in the sixties was the same as his share of mercury. In other words, between 1661 and 1669 his refineries were probably responsible for between a seventh and an eighth of Potosí's silver output. The mercury figures also indicate, as would be expected, a distinct growth in his production over the decade. In the years from 1661 to 1665, he bought 9.7 percent of the quicksilver entering the Treasury; but in the four years from 1666 to 1669 he took 18.6 percent. So it is probable that by the late sixties he had become the source of almost a fifth of Potosí's silver output.[48]

The late 1660s seem also to have been a time when he took on more refining plant—as would be expected in view of the growth of his mine holdings and of his mercury purchases. He had to have mills in which to put that mercury to work producing silver. But it is not easy to follow his acquisition of refineries precisely. A listing of *azogueros* in Potosí in mid-1665 shows López as operating only one stamp mill (*cabeza de ingenio*).[49] But various other pieces of evidence suggest that he had more than that at his disposal by the middle of the decade. From December 1663 he can be seen renting from the Treasury an *ingenio* belonging to don Pedro de Brizuela that had been seized for debt.[50] Another passing reference shows that by the end of 1664 he had inherited a refinery from one Juan de Lázaro, and had bought another from don Juan Marín Garcés.[51] And by June 1665 he was, it would seem, engaged in buying the refinery previously belonging to Bartolomé de Uceda, which he had operated through an administrator for several years past. Again, from August 1667 he rented from the Treasury another refinery, this one confiscated for debt from a well-known refiner in Potosí, don Pedro de Yebra Pimentel.[52] By the end of the sixties, therefore, he may have had as many as five refineries in the Villa Imperial at his disposal, possibly controlling a twelfth of the refining capacity of Potosí. In the new decade he continued to lease plant from the Treasury. The refinery of Yebra Pimentel, for instance, remained under his control until at least 1675, and in 1671 he began renting one seized for debt from a man named Gervasio Navarro.[53] Both these were in Potosí, but one other lease was granted to him in 1674 of a refinery out in the district, at San Antonio in los Lipes. This signals the growth of his mining interests outside Potosí in the 1670s, which indeed was to be remarkably vigorous.[54]

These, then, are the bare elements of the mining and refining apparatus that Antonio López de Quiroga constructed during the 1660s, his first decade as predominantly a maker, rather than a

broker, of silver. His energies were concentrated on the Hill of
Potosí itself, and especially on the Amoladeras zone on its western flank. The obviously large volumes of high grade ore that he
extracted from this zone, through the array of adits that he developed to get at his ores more easily, fed his growing refining plant
to such good effect that by the end of the decade he was by far the
leading silver producer of Potosí. Right at the end of the decade,
he began to turn his attention also to other sections of the Hill,
beginning with a sector called "Berrío" after a successful sixteenth
century miner. This was a part of the eastern slope of the Hill
that had once given great wealth, and in 1669 López went into
partnership with Jofre Ibáñez de Arreguia, who had long worked
an adit there. The gallery had clearly not been well kept up, and
López undertook to provide labor to clean it out in return for a
quarter claim in all the veins that Ibáñez had cut with his adit,
and a half share in any mines he might register in new veins discovered from then on.[55] In due course, this agreement would
yield its fruit. And so would mining work outside Potosí that
López began to tackle at about the same time as he set himself
up as a partner in the Berrío adit. But in the unfolding of his career,
the 1660s, his fifth decade of life, must be seen as belonging to
las Amoladeras. There he experienced his first, and crucially successful, venture into mining.

The physical size and the technical originality of that venture
are easy enough to discern, and impressive they certainly appear.
What may have been just as remarkable, however, but much more
difficult to make out, are the business procedures that López
characteristically used. Only the general strategy seems clear:
whenever possible he strove to acquire the assets of silver production—mines and refineries—at minimum cost. He did not engage in grand, new explorations for ore, but rather subjected known
deposits to more systematic exploitation than they had received
before. He built no new refineries in the 1650s or 1660s, but
rather preferred to take advantage of existing plant, either through
administrators, or through leases of *ingenios* seized for debt by
the Treasury. Obviously enough, he sought by these procedures
to limit his capital outlays as far as possible, and to use the

fixed capital already in place as effectively as he could. It might be argued that circumstances particularly favored him in this strategy—the circumstances of decline in the mining business of Potosí during the fifties and sixties. In such times of contraction, assets such as refineries were, it could be reasoned, available cheap. This may well be so; but mining had been in a similar condition of contraction for many decades, and in all that time no figure comparable to López had appeared. The opportunity may have been at hand, but there had been no one to seize it.

Of the more detailed aspects of his business administration, López left little trace. Clearly by the end of the decade, supervision and coordination of so many mines and refineries must have been a complex matter. Keeping track of multiple and convoluted transactions was something that López would have learned as an *aviador* and silver broker. And in both capacities he would have gained some familiarity with at least the refining end of silver production. But his knowledge of *mining*, in the specific sense of ore extraction, at the beginning can only have been sketchy; and in truth there is no evidence that he himself ever directly administered a mine, or for that matter a refinery. His practice was to place salaried administrators in underground workings and *ingenios*, and, at least in some cases, allow them broad freedom of action. This was rather common procedure in Potosí. Most *azogueros* who possessed mines, for instance, employed a *minero* to operate them and administer a work force; similarly, refineries were often put in the charge of a steward, or *mayordomo*. López did the same. Where he may have differed, as far as the scant evidence permits a view, is in being careful, or skillful, enough to choose only the best managers, and in rewarding them unusually well. His principal *mayordomo* of the sixties, for instance, was a man named Ambrosio Ruiz de Villodas, who in 1672 went on to become one of official inspectors, or *veedores*, of mines in the Hill, and in due course, in the eighties, an *azoguero* and mine owner in his own right. He was obviously a knowledgeable man in many aspects of silver making and an active one besides. And on the question of wages, Arzáns de Orsúa y Vela recalls that López paid his managers well, or allowed them to pay themselves well.

All the Spaniards who served the *maestre de campo* Quiroga for four, five, or six years came away with healthy fortunes, and in advancing him, themselves grew rich. Many of them, besides receiving a most considerable salary, stole the best of his ores. When Quiroga was told this [about one of them], he said with great calm: "Let him steal, for so he will work and will bring me greater profit from the mine on account of what he takes for himself.[57]

Such complacency, such veritable mellowness, in fact, hardly seems to fit with the forceful impatience apparent in the Antonio López who rose so dramatically during the 1660s to dominate silver production in Potosí. And possibly the attitude is really one more typical of López's later years, when Arzáns as a young man might have known him, or known much about him, than of his early or middle career. But the leniency could also be read as expressing confidence—López's confidence in his judgement of men and their capabilities, or a broader confidence, born of wealth and success, that all but the greatest adversity could be handled without difficulty. Sureness of self is certainly present in López throughout his career. And perhaps it was that intangible but powerful quality that, more than any particular technique of administration, carried him through his broad and complex undertakings to success. The venture at las Amoladeras in the Rich Hill in the 1660s certainly seems the fruit of such self-certainty, and so do the several projects of equal magnitude that López would launch at sites across the entire Potosí district during the seventies.

3

WIDER HORIZONS

MINES TO THE NORTH

THERE IS NO HINT, INDEED, OF COMFORTABLENESS IN THE ANTONIO López de Quiroga that manuscripts from the 1670s reveal. On the contrary, they show little in him but force, ambition, and impatience. By the beginning of this new decade, Potosí alone seems no longer to have been a spacious enough arena for López's energy. Having triumphed there in the 1660s, he now sought new fields in which to test his abilities and his fortune. It is true that he went on working the Hill of Potosí to the end of his life, with increasingly ambitious schemes, and with, apparently, continued success. But in mining his attention now turned more and more to sites out around the vast highland district of which Potosí was the economic center. And from his base in the Villa Imperial, López's gaze traveled even farther, over the eastern ranges and foothills of the Andes and into unexplored hills rising from the rain forests of southern Amazonia. There he hoped to find new sources of precious metals, and also souls to be converted to the Faith. To discover either one or the other, but preferably both, would bring him fame beyond what he could hope for from his enterprises in the mountains. Nor did his vision stop there. Beyond and better than any recognition that South America and colonial society could grant him would be acknowledgment of his achievements from Spain. For someone who had done as much for the monarchy as he had, he thought, an aristocratic title would be a proper reward. So in the early 1670s he set off in determined pursuit of a title, firing off to Madrid a string of glowing reports about his services to the crown. It was undoubtedly one of the great disappointments of his life that the king and his ministers never saw fit to make him the Count or Marquis he felt he deserved to be.

Local recognition he did get, though, in growing measure. In 1668 he appears as one of the representatives (*diputados*) of the *gremio* of *azogueros* in Potosí—the loosely organized guild of silver producers whose main activity was to lobby on their own behalf before the various governmental bodies whose decisions affected mining. In August 1668 López, with three other deputies, signed a letter to the crown praising the administration of a recent *corregidor* of Potosí, don Juan Jiménez Lobatón, for his favorable policies towards mining.[1] It must have given him keen satisfaction, as a man who had only recently become a silver producer, to find himself addressing his monarch as a representative of such an economically important group as the *azogueros* of Potosí were. For, even in decline, Potosí was still the major silver center of the empire, and the members of its mining *gremio* still considered themselves, and rightly enough, key men in the well-being of the Spanish state.

Within a very few weeks of signing that letter, López had a meeting that can only have gratified yet further his desire to be recognized. It was a meeting, too, that did much to set him off on new tracks of enterprise that held out prospects of fame. The encounter was with the newly-arrived viceroy of Peru, don Pedro Fernández de Castro y Andrade, tenth Count of Lemos—a man whose family, with its ancestral seat at Monforte de Lemos in Galicia, had had long contact with the Quirogas, though it is impossible to tell from López's account of the meeting whether the two had any earlier acquaintance. Arzáns de Orsúa y Vela certainly suggests some greater informality between Lemos and López than would be expected from a mere sharing of region of birth. "He gave him a slap on the back, saying 'You're a rich fellow now, Antonio,' and then, wishing to add to this intimacy with greater liberality, said 'You should know that the Countess, your countrywoman, is expecting a child, and I wish you to be a godparent.

What do you say to that?' "[2] As always with Arzáns's stories, and particularly those including reported speech, scepticism is in order here. But there is enough circumstantial evidence about this meeting to give some plausibility to the report. López and Lemos did meet, without a doubt, and the Countess of Lemos did give birth to a child in 1668, though certainly a little before the meeting between her husband and López took place.[3] Arzáns has López de Quiroga accepting the viceroy's flattering offer with gratitude—to the tune of 50,000 pesos for baby clothes.[4]

Antonio López met the viceroy at Puno, on the eastern shore of Lake Titicaca, in September or early October of 1668. This is the only long journey away from Potosí that López is recorded to have made during his fifty years in the Villa Imperial. Few viceroys, either, for that matter, ventured far from their capital in Lima, and the last one to have come anywhere near Puno was don Francisco de Toledo, almost a hundred years before. What had drawn Lemos to this remote place was a severe and persistent civil disorder in a mining camp called Laicacota, a league away from Puno, where rich silver ore had been found in 1657. Fortune hunters rushed in, and a brief burst of prosperity followed. But throughout the 1660s factionalism, brought on by economic rivalries, enmity among regional groups from Spain, and tensions between Spaniards and colonials, had plagued Laicacota, to the extent that royal control was lost. Lemos made it his first important item of business to quell the disturbances after his arrival in Peru in September 1667.[5] He reached Laicacota on August 3, 1668, and stayed there until October 13, reestablishing crown control with harsh and peremptory measures, among which was the razing of the mining settlement of Laicacota, which reportedly consisted of over 2,000 buildings.[6] He then left for Cuzco, to deal with minor civil disruptions there.

López's telling of the meeting with Lemos, written five years later, while not suggesting any particularly close relationship between the two of them, does certainly give the impression that he considered himself the sort of person who was entitled to deal with viceroys on equal terms. Part of that assumption doubtless came from the fact that he went to this encounter as, once again, a *diputado* of the silver producers' guild in Potosí—though, perhaps the better to convey a sense of his own importance, he neglects to mention that here; instead, he makes it appear as if his desire to have weighty discussions with Lemos sprung directly from his sense of duty as a leading citizen of the colony.[7]

I went to see him, disregarding inconveniences and the distance of 140 leagues [it is actually about 330 miles as the crow flies], since I wished to discuss with him important matters for the better conduct of the business of the Kingdom. The entire undertaking occupied me for twenty-one days. I especially begged him to come to Potosí to see to the administration of the *mita*, which is in great decline and badly in need of reform. And he had resolved to make the journey; but then news arrived of disturbances in Cuzco. He went there, and found the trouble to be that a number of bandits [*forajidos*] had killed fourteen Spaniards. Immediately after that followed the sack of Porto Belo, and the viceroy had to go down to Lima to organize support, charging me as he did to so work the large adit in Laicacota, which had been left abandoned. . . .[8]

It would indeed have been a coup for López to have brought Lemos to Potosí, which no viceroy except Toledo had ever visited, or would ever visit. If he had managed that, however, and, as a great silver producer himself and a good representative of the *azogueros*, had urged Lemos to take steps to increase the delivery of forced laborers to Potosí in the *mita*, he would have met with disappointment and perhaps a cooling of friendship. For Lemos proved to be no ally on that issue. Perhaps as a result of his first-hand acquaintance with highland Indians during his journey to Puno and Cuzco, he soon revealed himself as a determined foe of the *mita* of Potosí, and actually went as far as proposing to the crown in 1670 that it should be abolished, to spare native laborers the suffering and abuses that it brought them. Nothing came of this proposal, partly because of Lemos's death at the end of 1672, and partly because the home government feared a decline in silver output if forced Indian labor were withdrawn from Potosí. Antonio López, whatever his sense of comradeship with this countryman in high places, may well have felt some relief at the viceroy's death, because with it the threat to the *mita* was reduced.[9]

He must have found it far easier to follow Lemos's urging that he should set to work renovating the great adit (*socavón real*) of Laicacota than to sympathize with his views on forced Indian labor. Indications are that he plunged into that task in 1669, sending up to Laicacota an agent named Domingo Vázquez del Villar with an initial sum of 30,000 pesos to start restoration of the adit.[10] In February of 1672, Lemos adjudicated to López, perhaps in recognition of his efforts up to then at Laicacota, another adit and adjacent mines, which for reasons unstated had been confiscated some years before from one don Antonio de Andrade.[11] But despite this addition to his opportunities on the pleasant shores of Titicaca, this most northerly and distant of López's enterprises never seems to have amounted to much. In 1674 he announced that he had stopped work there because of renewed violence among the "bandits" whom Lemos had suppressed several years before. This, on the other hand, seems to have been only a temporary interruption, since at the end of the 1670s, the *corregidor* of Puno reported that López's efforts with adits had drained various mines, and that new wealth from Laicacota was in the offing. Nothing further, however, developed from this. And in 1689 López complained of having spent more than 70,000 pesos in the place for no return.[12]

More rewarding, though not startlingly so, was another venture outside Potosí that Antonio López took up at about the same time as his attempt at renovating Laicacota. This was a project much closer to home—at Porco, a silver mining settlement only twenty-one miles southwest across the mountains from the Villa Imperial, though a far more ancient source of silver. Native miners had dug here long before Europeans came onto the Andean scene, and Spanish mining activities had started in 1538, seven years before the discovery of the Rich Hill of Potosí. For the rest of the sixteenth century, Porco had been consistently worked; but in the seventeenth, mining was sporadic there, and López was later said to have got work started again after many years of inactivity.[13]

It was probably in 1669 that he resolved to attack the principal silver deposits of Porco, which were badly flooded. This, in contrast with his effort at Laicacota, was an undertaking purely of his own choice. It is likely that encouraging experience with adits at the Amoladeras site in Potosí lay behind his decision to use a similar technique at Porco, with a low adit there that would drain and improve access to mines that could no longer be profitably worked. The plan took much longer to realize than he expected, and its final outcome is hard to judge. But clearly this adit grew into an engineering project far more ambitious than the work at Laicacota. In 1674 López reported that because of the hardness of the local rock five years' work on the gallery had advanced it a mere 500 yards. And even that progress had been accomplished only through the expenditure of over 30,000 pounds of blasting powder. He urged himself on with, "the hope of immense wealth that, according to the tradition coming down from ancient miners, don Francisco Pizarro took from these mines."[14]

The Pizarros—Francisco and in particular his younger brother Hernando—had indeed done well from Porco around 1540. But while the thought of working the same mines as had filled the coffers of those great and almost mythically distant *conquistadores* may have inspired López in his long attack on Porco, no useful result seems to have come from his unceasing efforts until July 1681, by which time he had spent over 300,000 pesos on the project. In that month a ventilation shaft (*lumbrera*) cut upwards from the adit ran into reportedly good and plentiful deposits of ore, and López's prospects at Porco seemed good.[15] But eight years later, work was still in progress. The adit by then had grown to a length of 1,100 yards; and witnesses who had seen it and toiled in it convey no impression of either its being finished or having brought López notable returns.[16] The broad opinion seemed to be that, although the veins cut by the gallery had not turned out to be as rich as hoped, López's scheme had been generally useful in simplifying access to mines in Porco (see fig. 3). And in doing that, it can only have tended to promote silver production there, though not to any spectacular degree.

In one respect, nonetheless, the adit that Antonio López made

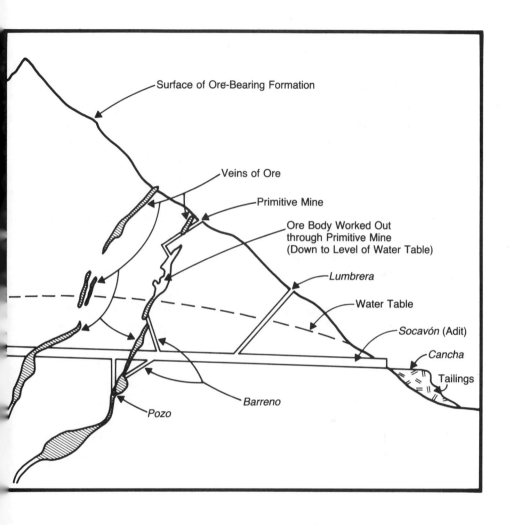

Cross Section of Mine Workings (Not to Scale)

at Porco was truly spectacular, and possibly even a revolutionary departure in Spanish American silver mining. For this seems to have been the first undertaking, in either Mexico or Spanish South America, in which blasting was employed in the pursuit of silver. Earlier use cannot be wholly ruled out, especially since powder had reportedly been employed as early as the 1630s in the Huancavelica mercury mine, to the southeast of Lima, precisely with the aim of pushing an adit forward faster than could be done by hand. But in Potosí and its surroundings, no hint of blasting is to be found, not even in the Amoladeras works of the 1660s, until López boasts, in 1674, of having turned it to advantage in the Porco adit, which had been begun at the end of the sixties. Neither is there any report that blasting was used elsewhere in Andean silver mining before that date; and in Mexico (as in Spain itself), the technique seems not to have been adopted until after 1700. But in Potosí, from the time of López's adit at Porco onwards, blasting seems to be taken as a commonplace technique in mining generally.[17]

To make even rough numerical estimates of the benefits that the new technique might have brought is impossible, given the scarcity of information on mining costs at the time in Potosí and the inadequacy of what little information of that sort is available. But it certainly seems reasonable to propose that blasting might have helped miners enormously in two particular ways. The first was to make the construction of long subterranean galleries a far more attractive proposition than before, since gunpowder would enable rock to be shattered much more quickly than could be done with hand tools. "The hardest rock that is may be broken with gunpowder, by observing in the first place a conveniency for the piching of your hole," noted a skilled English miner seventy years or so after López started using powder at Porco.[18] Long adits, therefore, such as Antonio López liked to drive, may well have become significantly cheaper to make than before. And that may go some way toward explaining López's liking for them as a means of giving a new lease of life to old and abandoned mines. Second, blasting, by providing a means of advancing at higher speed than previously possible through the rock in which the ores were encased, may have brought considerable economies also in the miner's crucial task of finding new ore bodies in which to work. In both cases the savings resulting from a lesser need for actual ore cutters may have been offset by a greater call for men to carry rubble out of the working. But this task itself could be far more efficiently performed when a gently inclined adit was the route by which waste was removed, instead of a series of steep and unstable ladders leading up several hundred feet to the surface. All considered, then, blasting seems likely to have brought great gains to silver miners through reducing the costs of extraction and exploration (not to mention the other benefits of improved drainage and ventilation through the adits that had been made with the help of powder); and Antonio López seems to have been the first silver producer in Spanish America to have seen the potential of the technique, and put it into practice.

In addition to the two adits at Porco and Laicacota, López had four others out in the district under construction by the end of 1672. The most ambitious of them—in fact, the most ambitious venture that López ever carried out—was far off to the southwest of Potosí, at San Antonio del Nuevo Mundo in the province of Lipes. The other three were to the north of the Villa Imperial, in mining centers among the mountains of the province of Chayanta: Aullagas and Ocurí, at some fifty miles distance, and Titiri, the location of which is unclear.[19]

The least active and productive of these three enterprises seems to have been the one at Aullagas, though that was the site among the three with which López had earliest contact. In 1657, when he was a fledgling *mercader de plata*, he and one of his companions in Potosí in the same business, Antonio de Cea, claimed to have discovered the silver ores of Aullagas. What they seem to have meant is that they had financed the miners who were then working those ores. Whether the discovery really was new cannot be said for sure. Although there are no references in documents to Aullagas before the mid-1650s, some mining apparently took place there before that, since in 1656 the Treasury officials of Potosí expressed a hope that with the drainage works then under way at Aullagas, silver production in the Potosí district would rise.[20] Some workings, then, must already have existed to be drained. Certainly, though, they had not been of much consequence. Nor, apparently, did they amount to much later, despite expectations in the Potosí Treasury. In 1680, the president of the high court of La Plata described Aullagas as a *mineralillo*, or minor ore deposit, which was doing quite well at the time, considering its small scale (*poca gruesa*).[21] Antonio López reports having an adit under way at Aullagas at the beginning of the 1670s, but he makes no reference to it in an account of his services (including a listing of his adits) submitted in 1679.[22] Possibly, then, he lost interest in the Aullagas mines in the course of the decade—when he was certainly at work in places that offered greater rewards.

Ocurí, seventeen miles or so southeast of Aullagas, and a center of silver production since around 1649, was one of those places. Here López invested quickly and heavily after 1670. By the middle of 1674 he had not only mines there, but also a refinery, full ownership of one adit, and a half share in another with a business partner named Martín de Narvaja. Narvaja was clearly the originator of this second adit, and López's role in the seventies was to supply credit for its continuation. By his own account his contributions had allowed the adit to be lengthened by some 275 yards by mid-1675.[23]

Many of what became López's standard operating practices can be seen in action at Ocurí in the early seventies. As far as can be told, he never went there, preferring to administer his refinery and underground works from Potosí through correspondence with a trusted and capable agent. One of the agents he used at Ocurí, Antonio Largáñez, had, in addition, another quality that López increasingly preferred his managers to possess: local administrative power. Largáñez, who became López's employee in 1674, was also the lieutenant *corregidor* of Ocurí and Aullagas, and, as such, he trod very close to the limits of the law in accepting a private job from López. He was sufficiently aware of this conflict to refuse to make mining claims for López—since it was the task precisely of a *corregidor* or his lieutenant to register and approve such claims. But he did agree to use his knowledge of Ocurí to promote López's business there and to supervise operations of his mines, adits, and refinery.[24] And his administrative authority can only have worked to his master's benefit.

Though he ran his ventures at Ocurí by remote control, López evidently kept close track of conditions and progress there. Less than a month after he appointed Largáñez as his administrator, he noted in a letter of instructions that the stamp shoes of the crushing apparatus in the refinery were small, and would therefore mill ore slowly, "and so your worship should be advised to send for stamp shoes, and mercury also, if necessary."[25] López evidently took trouble to keep the refinery, and the underground workings, well supplied with tools and materials, and this care is confirmed by later approving comments on how completely equipped his refining plants generally were. Between July of 1674 and September of 1677, Largáñez received from him 107,000 pesos' worth of *avío* for the production of silver at Ocurí, of which he kept careful track: cash, mercury, stamp shoes, crowbars, wedges, hoes, saws, adzes, sieves, scales, weights, nails, planks, padlocks, steel, tin, copper, tallow, candle wick, grease, various types of cloth, and coca. In those three years, the refinery, which was rebuilt during Largáñez's stewardship at a cost of 14,142 pesos, produced some 5,850 pounds of silver, worth 76,772 pesos.[26]

Also typical of López's business practice was his repeated insistence that profits should be immediately reinvested in means of producing silver. "Anything that the refining mill yields shall go toward the expense of the adit . . . ," he commanded Largáñez in July 1674.[27] Two years later, he urged that any ores in the mill should be refined with all speed so as to provide silver for *avío*, especially for the adit, which should be driven forward as fast as possible.[28] This insistence on reinvesting profit was to be remarked on in later years as one of López's most distinctive qualities. And finally, noteworthy among his practices at Ocurí, is his use of blasting. No evidence exists one way or the other of his having put this new technique to work at Aullagas; but at Ocurí he certainly did so by or in 1674, since he informed Largáñez in August of that year: "The Indian [who carries the letter] is bringing the powder and the crowbar that your worship requests."[29] There is no hint, unfortunately, of how extensive a use López made of powder in driving his adits at Ocurí.

The base that López laid down at Ocurí in the 1670s was evidently a solid one, since in 1689 he was still fully active there. In fact, two of the four refineries that he was then operating outside the Villa Imperial were at Ocurí (the others being at Titiri, nearby, and at San Antonio in los Lipes), and were described as being "of splendid construction."[30] His two adits, which he said had cost him in excess of 200,000 pesos to cut, were still functioning—and obviously functioning to good effect, if they were giving access to mines that provided enough ore to occupy two refineries. Supervising the entire enterprise at Ocurí by this time was the *maestre de campo* don Juan Antonio de Bóveda, one of López's brothers-in-law. This, too, is a characteristic note in the orchestration of his affairs in his late career. Trust was increasingly placed in family members, and less in purely business partners, as time went by.[31]

Another lasting success, and one displaying once more Antonio López's characteristic modes of procedure, was Titiri. Like Ocurí, this settlement dated from the late 1640s.[32] López began to pay attention to it around 1670, at the same time as he started work in Ocurí and Aullagas. His particular aim was to exploit a vein of silver ore named Santo Domingo Soriano, which had been attacked right after Titiri was first opened up, but then abandoned about 1650. The working had later suffered internal collapses that blocked access to the ores, and flooding added further difficulties. López set about having the mines in Santo Domingo cleared in 1670; but the fallen rubble proved such an obstacle that in the following year he decided, true to his by now established form, that an adit aimed at the lowest existing workings on the vein would be a more profitable approach. In 1671 he petitioned for, and was granted, a rudimentary tunnel that had been started around 1650, but abandoned after only 8 yards on account of the hardness of the rock. Work started on the extension of this gallery in mid-1671. By mid-1675 it had progressed another 80 yards, cutting two new veins along the way; and by October 1675 it had grown to a length of 160 yards, and had recently intersected not only the old and abandoned workings in the vein Santo Domingo Soriano, but also, 7 yards farther on, another vein in which López's manager was quick to stake a claim. The total cost of the gallery to that point was over 30,000 pesos.

Evidence is lacking to show that López's miners used blasting in excavating the adit at Titiri, though it is likely enough that they did, given the reported hardness of the rock. But they certainly used powder in attempting to open up the old workings in Santo Domingo Soriano, blasting out new ventilation shafts and narrow communication passages (barrenos) in rock that, again, was reported as being very hard. This, in fact, is the first definite instance in the Potosí district of powder's use in a *mine*, as opposed to an adit. López's underground explorations and improvements at Titiri evidently yielded rich ore. One report dates his extraction of "*mucho metal bueno*" there back to 1672; and by mid-1675 he had a refinery running at Titiri, which suggests a large and steady supply of silver-bearing mineral. This mill was still in place and operating in 1690, although some doubt was expressed then about its profitability.[33] Nevertheless, all the available evidence points to López's having done well at Titiri over the years.

San Antonio del Nuevo Mundo

"There is not a mining center in the realm that he does not supply with credit and work on his own account," wrote an admirer of Antonio López's in 1690.[34] This was an exaggeration, though a pardonable one. By that time, López loomed so large in Potosí that it must have seemed that he had spread his tentacles to every place in Charcas (if not quite the entire kingdom of Peru) in which there was the merest glint of silver. In reality, López paid no attention to many sites around the Potosí district that had at some past time yielded substantial amounts of silver: Oruro, Chocaya, the Salinas de Garcimendoza, and Esmoraca, to mention just a few[35] (see fig. 4). His preference, it seems, was for places (with the exception of Porco and Potosí itself) where ore deposits had been found during his own time in Potosí, or at the earliest very shortly before his arrival. Nevertheless, it would be wrong to give

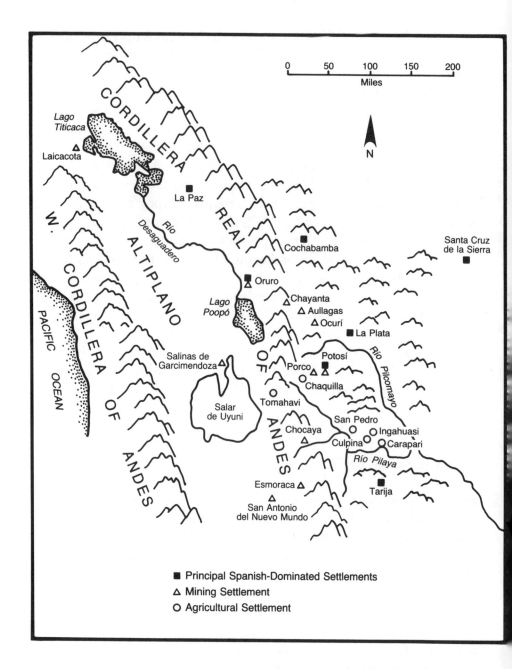

Major Towns of Charcas and Sites Associated with Antonio López de Quiroga

the impression that he made no effort at all with more venerable deposits, although his experience with the two that he did try does not seem to have been rewarding enough to have moved him to tackle others in the same category. These were Berenguela, in the province of Pacajes, and 280 miles northwest of Potosí; and Tomahavi, in a rugged mountain chain of the same name, some 65 miles southwest of the Villa Imperial. Both were sites of ores that had been worked, on and off, for many decades—Berenguela from what for miners of López's day was almost time immemorial in the mid-sixteenth century, and Tomahavi from about 1614.[36] In both places, López set about driving adits in the 1670s, but he makes only the most passing of reference to them in his later reports to the king, which suggests, as he was not hesitant in proclaiming the magnitude of his efforts, that these galleries came to little.[37]

The mines in Chayanta, then, whose discovery dated from the late 1640s, were far more characteristic of the deposits to which Antonio López apparently preferred to direct his attention. Perhaps, indeed, he thought that older mines would simply have been worked over too often and too thoroughly to retain any further substantial wealth. To take advantage of what was already discovered, but not too well explored, was perhaps his preferred line of approach. Aullagas, Ocurí, and Titiri all fitted into this category, and indeed may have added up to his most successful venture outside the Rich Hill of Potosí.[38]

For his contemporaries, though, López's most remarkable endeavor away from the Villa Imperial, and possibly the most spectacular of all his works, even including the complex of adits and mines at the Amoladeras, was not in Chayanta, but far away to the south in the province of los Lipes, at San Antonio del Nuevo Mundo. "A work unique in the entire kingdom" was the assessment made of the adit and linking galleries that he opened up here during the 1670s and 1680s.[39]

Hardy men from Potosí had been producing silver in los Lipes since the early 1570s, when ore was first found there at a place called Uslloque.[40] It was a large province, reckoned then to be almost a thousand miles in circumference, lying west and southwest of Potosí and taking in much of the southern *altiplano*—the great plain stretched 11,000 to 12,000 feet up between the coastal and inland ranges of the central Andes. Here the rivers flow inward from the mountains and volcanoes on both sides, creating in the wetter, summer months great sheets of shallow water where Andean flamingos feed and wade. In the dry winds of the winter the water evaporates, leaving crusted salt flats on the surface. Rarely does rain fall on the plain itself, for it is part of the mountain desert of the Atacama, high, cold, and desolate. If Potosí now seems remote and forbidding to the outsider, for Spaniards living there in colonial times the Villa Imperial, set among mountains and offering a way of life with European, as well as local, roots, seemed welcoming and familiar in comparison with los Lipes. That was a   bad and poorly populated" land, said a Treasury official from Potosí in 1581. In its whole vast area lived only some 5,000 Indians in very small and scattered villages: 4,000 of them Aymara speakers who raised llamas and alpacas, grew crops, smelted silver, and traded in Potosí; and a thousand others who were much poorer, living from hunting *guanacos* and *vicuñas* (each a type of wild Andean camelid), eating fish and roots. But in ores of silver and lead the province was rich, and the people took these minerals from veins unknown to the Spanish, and smelted them in their houses and in wind-blown furnaces set out on the hillsides.⁴¹

The vision of mineral wealth in los Lipes danced constantly before the Spaniards' eyes in succeeding decades. And, frequently enough to keep the vision bright, they found silver there, so that small mining settlements came and went. Especially did interest in los Lipes grow when the decline of the Rich Hill of Potosí became clearly inexorable in the seventeenth century. In 1635 the president of the high court of La Plata informed the king:

The province of los Lipes is one of the most unusual in these realms, because, in addition to there being in it most strange and varied things that Nature does not produce elsewhere, it is the best supplied in rich sources of silver . . . of any region that has been found. . . . It would be most useful to . . . examine all that it contains, and explore a part of it that lies between the road that goes to Tucumán and the coastal route that leads to Chile. This area is called in these parts the *Nuevo Mundo* [New World], not because of its size, but because it is unknown and hardly anyone has gone into it. Some years ago Luis del Castillo and four companions of his did enter there, and explored for seven leagues; but they quickly came back on account of some renegade Indians. He told me, though, (and he is a man of all truthfulness) that along their route they came across many very rich deposits of ore, and, to judge to everything that they could see, that land promises great discoveries. . . . Little is to be lost here, and much to be won.[42]

One of the things that was won in this *Nuevo Mundo*, a decade or so after this letter was despatched for Spain, was indeed a remarkably productive new mining deposit (though whether it was revealed by any such organized exploration as the president of La Plata recommended is not clear). This was San Antonio del Nuevo Mundo, which may well have yielded more silver in the second half of the seventeenth century than any other place in Charcas besides Potosí itself. Precisely when its ores were found, and began to be processed into silver, is impossible to say. But by the end of 1647 fifteen men were there who described themselves as being "interested in mines," and were busy complaining to superior authorities about the conduct of the principal magistrate in the province, Ignacio de Azurza, whom for reasons undisclosed they did not like.[43] A year later, the population was apparently still thriving, for in December 1648 a freighter in Potosí took on a contract to carry almost three tons of wheat flour from the Villa Imperial to San Antonio on his llama train—early evidence of what would become a stream of supplies and men flowing along that route.[44]

For several years San Antonio continued to flourish, as usually happened with new mining discoveries. For a while, ores could be mined effortlessly and cheaply at shallow depths, and often weathering had enriched the minerals that lay near the surface. These very qualities of accessibility and wealth, however, attracted crowds of prospectors. As the intensity of exploitation grew and these superficial deposits were removed, *mining* had to begin in earnest. Workings were driven deeper, which made getting the ore out of them a slower and more demanding business. More workers were needed, and labor costs edged upward as the mines went down. At the same time, the quality of ore might decline and, worst of all, the lower levels of the workings would reach the water table and begin to flood. Diminishing returns tended to exacerbate disputes among miners over claims, with at best litigation as a result and, at worst, violence. In the face of these varied adversities, many prospectors would begin to drift away and look for less crowded and more rewarding claims elsewhere.

This is the sort of sequence that San Antonio seems to have gone through in the 1650s. Its early and first peak of output may have been reached around 1655, when the Treasury officers of Potosí told the viceroy that the district mines had actually contributed more silver to the remittances of taxes and other royal income then being sent back to Spain than the mines of the Rich Hill itself.[45] No other district mine is reported as being strongly active at the time, so it is likely that San Antonio was responsible for this unusual, and possibly unprecedented, state of affairs. The contribution of the district mines to the total amount collected on the crown's behalf in Potosí dropped, however, over the decade after 1660.[46] This is probably partly because San Antonio had passed its peak by then.

Certainly there was flooding at San Antonio to hinder production of silver. For the past twenty years, so the Viceroy Count of Castellar wrote to the king in 1676, San Antonio had been in "total decline" for that reason.[47] That seems too drastic a statement, since flooding never led to an immediate, and rarely to a total, collapse in production at a mining settlement. But flooding certainly did afflict San Antonio in the 1660s, as Antonio López, in the grandiloquent tones that he so liked to use, reported in 1674:

Seeing that the province of los Lipes was the region most plentiful in precious metals in the entire district, and that for lack of people willing to spend money it was abandoned and its profitable fruits submerged, I resolved to drive an adit there, beginning that work on October 19 of the past year of 1672 . . ."[48]

So López set out to try his fortune with yet another mineral deposit dating from the 1640s. His line of approach this time was in one respect different from his previous procedure in the Chayanta mines, at Porco, or at las Amoladeras in Potosí for that matter. In all those other places he had worked alone as an individual investor, but for driving the San Antonio adit he chose to form a partnership. What lay behind this decision he did not explain. And in the end it proved an unrewarding experiment that led López, perhaps as a result of frustration and impatience, into prolonged litigation with his companion. Possibly, however, he saw the construction of his planned adit at San Antonio as such an ambitious undertaking that even he preferred to spread its costs and risk somewhat. And possibly, also, the fact that his energies and resources were widely extended over several mining sites by 1672 inclined him to seek an ally in tackling San Antonio del Nuevo Mundo.

López's partner was Captain don Alvaro Espinosa Patiño de Velasco, proprietary treasurer of the mint at Potosí by the time of his alliance with López. This office he had bought for the very large sum of 124,000 pesos. He was a native of the Potosí district—to be precise, of a valley called Oploca in the province of Chichas, south of Potosí, where he owned land—and son of a well-known miner in that same region.[49] Why López selected this man as a partner can only be surmised: he must have known him well through contact at the mint; and he perhaps thought that Espinosa's resources were equal to the challenge that San Antonio might present.

In July 1672 Espinosa went to San Antonio to make arrangements for initial work on the adit. Antonio López, truc to form, remained in Potosí. The first necessary step—obtaining permission to make the adit from the *corregidor* of Potosí, as the chief mining authority in the district—had already been completed. Now Espinosa wished to negotiate an agreement with the mine owners of San Antonio that would specify what reward the partners would receive if their adit succeeded in draining the flooded mines. For this, a meeting was held in the house of the *corregidor* of los Lipes, who was one Captain don Cristóbal de Quiroga y Osorio, probably a relative of Antonio López's and probably also a man with whom he had had business dealings in the 1660s.[50] Certainly by a few years later he was one of López's trusted associates.[51]

The contract that emerged was one that seems likely to have offered López and Espinosa great potential gains, though at considerable cost. The mine owners agreed that if the water level in their workings were lowered as a result of drainage through the projected adit, they would grant the partners a half share in the portion of the mines that was thereby opened up to exploitation; however, what was now above water would remain the exclusive property of the present owners. If on the other hand the mines were completely drained by the adit, then the partners were to receive a full half-share in every mine so benefited, from its outlet at the surface down to its lowest levels. The conventional payments that owners of adits received under the law were a fifth of the ore extracted from a mine that was worked through the adit, or a tenth from a mine that the adit merely drained.[52] These payments entailed no further effort, except maintenance, by the owners of the adit once it was finished. López and Espinosa, on the other hand, agreed to take not a half share of ore, but rather a half of each mine that their project might serve or revive. To take advantage of that share, they would have to put in labor to cut and remove ore—a further and considerable expense beyond the construction of the adit, which in itself was likely to be a most costly undertaking. So it would seem that the two partners were prepared to accept the prospect of higher costs in recovering their investment and securing a profitable return than was customary among makers of adits. This was presumably because they knew that San Antonio had previously yielded great wealth, and they seemed confident that their scheme to restore its prosperity would be so successful that those extra costs would be more than offset. In other words, they may have expected that drainage would open up such rich ores, and in such quantities, that the cost of extracting them would be far less than their yield in silver.

López chose a skilled miner named Alonso Ruiz, who had been working on the adit at Porco, to supervise the driving of the new gallery at San Antonio. Ruiz began recruiting Indian labor for the job on October 20, 1672 in Potosí—twenty-five men, initially, many of whom were given advances of sixty to seventy pesos as a lure to go to San Antonio, "a place detested by the Indians, and from which they used to flee as fast as they could," said Ruiz.[53] He got round this difficulty, he added, not only by giving advance payments, but also by choosing "Indians who had no place in the town, such as chairmakers, shoemakers, pickpockets, and thieves."[54]

This initial band of workers was sent off to San Antonio, via Porco and the province of Chichas, in the charge of a young servant (mozo) of Ruiz's. Each received about three pesos while en route—whether as wage or for food Ruiz does not say, though he did also provide them with sugar and sweetmeats (cajeta), perhaps as treatment of sickness, since such things were commonly taken by Indians as medicines. An Indian woman was also taken along to cook for the workers.[55] In San Antonio, a small settlement of huts, roofed with ichu straw, was built for the workers to live in, and also a chapel, for which a priest was engaged. From a passing reference in Ruiz's remarks on the making of the adit, it seems that López and Espinosa made a practice of sending rough clothing, coca, and other supplies to San Antonio. Whether the Indian laborers whom they employed were forced to buy these things at high prices, as often happened on isolated rural estates in colonial Spanish America, is not clear; but profits from the sales, however made, went to cover the cost of food for the work supervisors (the mineros), the priest, and for "anyone else who arrived at mealtimes, since this is a place that demands such a custom."[56] Ruiz was proud of his treatment of the native laborers: "When the men and their womenfolk pretended to be ill, they were given sugar and mate tea, and when they [really] were ill they received food and whatever else they needed, so that I treated them as if they were my children." Ruiz's accounts also show entries for senna, coconut root, "resin of Santa Cruz," and the services of a barber—all bought in the interests of the native workers' health.[57] It was by taking such care of them, he maintained, that he was able to push work on the adit ahead so fast.

That work began on December 5, 1672, and continued day and night, with the exception of Holy Thursdays and Good Fridays, until January 2, 1678—263 weeks in all. The total excavated length by that point was 1,622 *varas*, or some 1,487 yards. This included 134 yards of water channel (*rasgo*) in the exterior, then 73 yards of open cut (*tajo abierto*), 753 yards of tunneled adit (*socavón*) from the entrance to the work face, 341 yards in ventilation shafts (*lumbreras*), and 186 yards in four narrow communication passages (*barrenos*) linking the adit to mines in four of the major silver veins at San Antonio: La Concepción, the Veta Rica, San Juan Bautista, and Espinosa. Through these passages the workings in those four veins were drained by January 1678. Ruiz took great satisfaction in this accomplishment, and in the scale and quality of his work: "No other [adit] has been seen that is more perfect in levelness, direction, width, and height."[58] These last two measurements were indeed impressive: seven feet six inches by eleven to fourteen feet.[59] Not included in the height was a covered channel cut into the floor of the adit (and in the stretch of open cut), along which water draining from mines in the interior could escape without flooding the floor of the gallery. This channel (*larca*) was two feet deep by one foot five inches wide, and was paved over with flagstones. This paving may have been done not only to give more space for workers moving to and from the mines, but also to provide a rolling surface for carts bringing out ore and rubble. Certainly during the cutting of the adit, Ruiz spent 242 pesos and 4 reales for "a small cart . . . to carry out the rock that the pickmen loosened." Whether wheeled vehicles were still used later for moving ore out the mines is not revealed by subsequent documents. But even if the practice was restricted to the making of the adit itself, it was a small and potentially revolutionary step in mining technique in and around Potosí; for there is no sign before this that ore and rubble were removed in anything but the traditional way—in sacks on men's backs.[60]

The better to convey the scale of his accomplishment, Ruiz contrasted the size and rapidity of construction of his new adit with the history of the most famous adit driven into the Rich

Hill of Potosí in the sixteenth century. Using as his evidence Book 4, Chapter 8 of Acosta's *Natural and Moral History of the Indies*, he noted that this adit, known as the *socavón de Benino* after the miner who planned it and began the excavation, reached a length of only 230 yards in twenty-nine years of work between 1556 and 1585. Ruiz failed to mention, and perhaps did not know, that work on this project had been far from continuous, which made his comparison rather less telling. But he might well have agreed with and quoted the observation that Acosta was moved to make by the tardiness with which Benino's enterprise had advanced: "This goes to show how much effort men will make to go down into the bowels of the earth to seek silver."[61] Ruiz clearly thought that he, too, had made remarkable (and unusually successful) efforts in creating his great gallery at San Antonio.

In one respect, however, Ruiz and his men were obliged to expend less effort than Benino and others like him had had to put in during the sixteenth century. For Ruiz had the energy of gunpowder to replace much of the labor with pick, bar, and wedge that the "ancients," as miners of his day liked to call their early predecessors in Charcas, had inevitably had to perform. His tally of the powder used up to the end of December 1677 was 18,392 pounds, costing a total of 16,112 pesos and 2 reales. The use of this large amount of explosive had allowed him to vanquish, as he put it with pardonable exaggeration, the "invincible hardness" of the rock through which the adit was cut.[62] This had been the first "enemy" to challenge him at San Antonio. The second was flooding, even during the excavation of the adit. This he overcame "with new inventions of cartridges sealed with pitch to protect the powder,"[63] which allowed him to blast in wet rock (or perhaps underwater, though his wording is not conclusive on that point). Whose invention this was, he does not say either, though the implication clearly is that it was his own. Each cartridge held a charge of 6 ounces of powder; and each work team, or *barreta*, of three Indians doing the actual cutting of ore used two such charges, on average, in a twelve-hour day or night shift. So, calculating from the size of the charge and from the total weight of powder used, it seems that roughly 49,000 charges were fired up to January 1678. To break up each yard of rock in the 1,353 yards of tunneling or open cut completed by Ruiz, therefore, (ignoring the stretch of water-course outside the adit) would have required thirty-six charges (or about 13.5 pounds of powder).[64]

The completion of the adit (or more accurately, of what would prove to be its first stage) by the beginning of 1678 had very much the effect that López, Espinosa, and the mine owners must have hoped for: a lively recovery of silver output at San Antonio. As early as mid-March 1678, Viceroy Castellar told the king that the "socavón de los Lipes" had led to the draining and clearing of mines that had previously been quite useless. In the space of only a few months before February 1678, when the most recent remittance of royal funds to Spain had left the Treasury in Potosí, 18,000 pounds of refined silver had been received there from San Antonio, yielding, at the usual royalty rate of a fifth, over 60,000 pesos. This resurgence was, he said, the cause of great relief thoughout Peru.[65] Tax records at Potosí give persuasive backing to the viceroy's report, although the fact that they do not distinguish among different sources of silver out in the district makes it impossible to be absolutely sure of what was happening. In 1676, all the district mines together sent only about 5,000 pounds of silver to be registered and taxed at the Treasury in Potosí. In 1677, during which mine drainage through the adit at San Antonio began, total receipts from the district mines were some 7,200 pounds. And in 1678, when drainage of mines on four of the major veins at San Antonio was in full progress, or perhaps completed, the amount of silver coming into Potosí for taxation from the district was 40,430 pounds—an almost six-fold rise over the previous year.[66] Since there is no hint of any other new or increased mining activity anywhere else in the district in those three years, it seems extremely likely that revival at San Antonio del Nuevo Mundo was responsible, solely or largely, for this rapid increase in registration of district silver at the Treasury.

The gain proved lasting, as far as can be told from the recorded figures. Even though the president of the high court of La Plata was rather disdainful of San Antonio when he wrote to the king in June of 1680—"The Lipes [mines] are of some help, but not

everything that we had hoped for while the adit was being made"
—nonetheless registrations of district silver in Potosí rose almost
without interruption until 1683, reaching at that point 79,000
pounds, or 38.4 percent of the total production from Potosí and
the district combined. And even after that, district registrations
generally stayed around 40,000 to 50,000 pounds annually until
almost the end of the century.[67] Inevitably the amount contributed by San Antonio to these totals would have varied with the
passage of time. But evidence points to its having remained the
principal mining center in the district outside Potosí in the eighties and nineties.

As the adit drew water from the mines of San Antonio, López
and Espinosa would have been able to take up the shares in those
workings that their agreement with the owners had specified they
should receive. Only the sketchiest references can be found, however, to their mining and refining activities. In 1671, even before
the adit was begun, López had sent native workers in San Antonio to extract ore,[68] and presumably added to these once the flooding started to subside. In August 1674 he took a ten-year lease on
a refinery in San Antonio, once the property of the Ignacio de
Azurza who had displeased the miners there in the late 1640s,
for the almost nominal rent of 100 pesos a year.[69] The lease was
still in effect at the same rent in 1681,[70] so it seems probable that
López had used the mill over the intervening years. To his production of silver in it, however, no clue exists.

Of the costs he bore in driving the great adit, on the other hand
some record remains, largely because before the decade was out
he had plunged into legal disputes with both his partner, don
Alvaro Espinosa Patiño, and his manager, Alonso Ruiz, over what
they and he had put into the venture. In the course of the argu-
ments, the parties put forward as evidence various statements of
what the adit had cost, at least up to early 1678. The most complete accounting came from Ruiz, as might be expected, since he
was the manager of the project. With minor objections, López
seems to have accepted this statement of costs: from December
5, 1672 to December 31, 1677, 217,748 pesos and 2 reales had
gone into making the adit and its adjacent tunnels, shafts, open

cut, and water course.[71] The sum is on a par with López's spending on other adits during the early 1670s, although the work at San Antonio was probably more extensive than at the others and may have yielded a better return over the remaining two decades of López's life.

In the lawsuits that developed over contributions to the enterprise, it is difficult to tell who had right on his side, although the cases do show, if any further demonstration were needed, López's doggedness in pursuing his own interests. Alonso Ruiz had two complaints against López: that they had made a verbal agreement at the start that Ruiz would receive 20,000 pesos for his work on the adit, and also (rather contradictorily) that he was not just an employee of López's, but a partner in the venture at San Antonio. López denied both these claims. His attorney, one don Diego de Figueroa, dismissed the alleged verbal agreement as *"fantasías,"* and lodged a counter-claim to the effect that Ruiz owed López 50,000 pesos for unspecified "supplies" given to him while the adit was being made. Ruiz, so Figueroa (and presumably López behind him) asserted, wanted to be in partnership with López in every mining center in Charcas without contributing a single peso to any such arrangement. Everyone, indeed, knew that he could not do so because he was poor, and the situation was the opposite of what he had said: far from being owed anything by López, Ruiz should acknowledge the indispensable support that he and his family had received from his employer.[72] In 1679, don Cristóbal de Quiroga y Osorio, now no longer *corregidor* of Lipes, but acting as López's agent there, accused Ruiz of having "maliciously" misidentified the veins that the adit had cut so that he could register mines for himself in them (perhaps by claiming that they were previously unknown veins to which nobody could have a prior right). This seems to have been a confused charge and no more came of it. In the end Antonio López prevailed, though not as fully as he might have wished. In 1689, after many years of litigation, the high court of La Plata finally ruled, on appeal, that Ruiz owed López 18,560 pesos and 4 reales, and should pay up within ten days.[73] Perhaps he did so, for in the eighties he had apparently become a successful miner and partner in a refinery in Potosí itself, as well as the owner of a refinery in San Antonio valued at 26,000 pesos. In 1691, nonetheless, he still contended that he had been cheated out of his due as a partner in the San Antonio adit and grumbled on about López's "cunning tricks."[74]

The dispute between López and don Alvaro Espinosa Patiño seems clearer cut. The two partners had agreed to divide the costs of driving the adit equally, and Espinosa fell far short of paying his half. Of the almost 218,000 pesos that the first five years of work cost, Espinosa contributed only 40,000. In 1679, therefore, López lodged a claim against him for an additional 70,000 pesos. López also maintained that even those 40,000 had not come from Espinosa's own funds, but from two uncles of his, don Antonio Patiño de Velasco and don Fernando de Castro Guzmán, who had supplied goods and food to that value. A certain note of indignation and disdain comes into López's writing as he lays out his version of the story. Espinosa, he recalls, received 10,000 to 12,000 pesos a year from his treasurership of the mint in Potosí and, beyond that, was his income from his landholdings. But "he did not risk a single peso [of this] . . . in the company that we committed ourselves to for making the adit,"[75] nor would he now pay what he owed under the terms of the partnership. This recalcitrance, López thought, was not at all what Espinosa should be displaying. On the contrary, gratitude was the fitting sentiment, "for he should recognize that I, at the cost of my own estate, was in the process of making him one of the richest men in this realm, and he may still be so if he goes forward; but this willingness to lay out one's personal wealth is to be found in very few people."[76]

What Espinosa's reply to these contentions was (if he made one) has not survived. Despite López's complaints, the gains from the partnership seem to have remained as had originally been planned. The company was wound up in 1678, with the specified half of every mine that benefited from the adit being split equally between the two principals.[77] López never recovered everything that he claimed, and he had to wait until after Espinosa's death, early in 1687, for a settlement. In May of that year, he agreed to accept from the heirs of the estate 6,000 pesos in kind—100,000 pounds of jerked beef and 50,000 pounds of tallow made from cattle raised at Espinosa's estate at Oploca. Furthermore, all profits and gains from the adit at San Antonio were to go to López de Quiroga from this point onwards.[78]

Although it was generally thought that the company had reached its goal by 1678, in the sense that its adit had successfully drained and opened up rich mines in four of the major veins at San Antonio del Nuevo Mundo, López had hopes of still better things to come if the adit were to be driven yet farther forward. "The most substantial part of the place is still to be discovered," he wrote in 1679, "and so I am continuing the adit to cut another vein, the richest known in San Antonio, which contains native silver and is called Santo Domingo de Guzmán." The "geometrical measurements" that his miners had made indicated that the central gallery would have to be extended 210 yards for the intersection to be achieved; and the cost was likely to be over 100,000 pesos, since the rock was very hard, and powder would again have to be used. But this, he said, was "the greatest service that at present I can do for your Majesty, although the others I have performed in the past are very great, and are not to be found in any other of your subjects. . . ."[79]

Nevertheless, perhaps because he feared that Espinosa would continue to claim a share in any profitable outcome of the adit, López did not press ahead with it until the late eighties, after his partner's death. Perhaps, also, the sustained high level of silver production that San Antonio seems to have achieved by 1680 or so took the edge from his eagerness to increase his investment there. In any case, it was not until 1687 or 1688 that he again put people to work in extending the adit, carrying it to a total length of 1,470 yards by mid-1690. Although the Santo Domingo vein, with its great beckoning riches, was still beyond reach, he said another vein, called San Francisco, had been cut and much ore taken from it.[80] The wealth of this vein excited the enthusiasm of the president of the court in La Plata, who reported at about the same time that in the first eight months of 1690 San Antonio had yielded no less than 96,000 pounds of silver—which seems an improbable figure, since the total registered output of the district mines in 1690 was only some 45,700 pounds. [81] Perhaps the president had added in silver produced the previous year. Notwithstanding this exaggeration, San Antonio was probably producing well, as a result of López's further efforts, in 1690. It was, by one report, the most prosperous mining community in Charcas at the time.[82]

But beneath the surface, as it were, things were not everywhere so rosy in the San Antonio of the early 1690s as they were with the new deposits made accessible by the lengthening of the adit. Some of the workings opened up by the adit in the late seventies were again suffering from flooding, and had indeed been plagued by this ever-threatening problem for the best part of a decade. What evidently had happened was that after 1678, once the miners had excavated ore in the newly-drained sections of their working, they had pushed down into deposits that lay below the level of the adit. Flooding then inevitably returned, simply at greater depth than before. The adit still offered help in this situation, nonetheless. If water could be lifted up into it by some mechanical means, it could then flow normally down and out to the surface. This was obviously a more expensive proposition than letting gravity do all the work; but it was far cheaper and simpler than raising the water all the way to the mine entrance, which was the only means of drainage available where no adit existed. Well before the mid-1680s, water was being drawn up in this way from mines in the Veta Rica, the vein called Concepción, and another known as the Veta Ancha. A sump shaft (pozo) had indeed been dug by then to collect water, and a whim (torno) installed underground over it to lift water up and into the adit. This procedure eased the problem of flooding for a time. But clearly it increased the cost of extracting ore and was both less reliable and less able to handle large amounts of water than drainage by gravity through an adit. Mining in the three veins just mentioned, therefore, ground to a halt. Reports from local miners suggest that the Veta Rica and Concepción were out of action by 1685, and that little by little work tailed off in the Veta Ancha until there, too, it finally ceased in 1690 or 1691. Even the drainage sump was abandoned and its walls allowed to collapse.[83]

In 1693 López's general manager in San Antonio, don Joseph

de Ujúe y Gambarte, advanced a scheme to alleviate these difficulties and once again drain the veins intersected by the great
adit. He proposed to do this by installing yet another whim, or
perhaps two or three of them, to lift the water up to the adit. This
would have meant digging drainage sumps, and to make a start
in that direction he lodged a claim for the existing *pozo*, on the
grounds of its abandonment. To finance this scheme he suggested that anyone who had mines in the veins that his machines
would drain should contribute to the cost in proportion to his
potential benefit, and further, that anyone who would not contribute should have his mine workings declared forfeit and assigned to Antonio López. The high-handed nature of this proposal
did not sit well with many in the community of San Antonio,
though it seems to have been characteristic of Ujúe y Gambarte,
who, besides being López's administrator, was also the nephew
of Captain don Miguel de Gambarte, one of López's two sons-in-
law, and a very recent incumbent of the office of *corregidor* in
los Lipes to boot. So Ujúe had both López's force and considerable civic authority to bring to bear in pursuing his scheme. Nevertheless, it does not seem to have come to anything, perhaps
because it seemed, as it indeed was, merely a stopgap remedy for
the difficulty, and perhaps because it met such opposition from
the mine owners who were expected to pay for it, or, if they would
not, lose their workings.[84]

Registered silver output in the district mines dropped away
sharply after 1696[85]—perhaps, it might be suggested, as a result
of the final decline of San Antonio del Nuevo Mundo. Antonio
López's extension of his adit in the late 1680s may have provided
enough of a stimulus to sustain production for a few more years;
but once its good effects wore off, the general and prevailing decay
of the workings was revealed to its full extent. In 1703 a newly-
appointed *corregidor* of los Lipes reported gloomily that most
of the mines at San Antonio were "imposibilitadas"—unworkable
for one reason or another. With much effort he had managed to
drain one vein. But the effort seems to have had no dramatic
effect.[86]

The details of San Antonio's severe decline in the 1690s are still conjectural. But little doubt exists about the reality of its downturn or about its broad causes. The place had reverted to the state in which Antonio López had found it twenty or more years before—its ore deposits worked down to the levels at which flooding began, its mines therefore abandoned, and because abandoned, constantly deteriorating through collapses of rock and consequent accumulation of rubble. The promise of wealth was still there, for, apart from whatever lay below water, the vaunted riches of the Santo Domingo de Guzmán vein seem never to have been tapped by López's extension of his adit. But what was truly needed, instead of makeshift expedients of whims and wells, was another adit, driven lower yet to intersect the ore bodies below the level to which López's first adit had reduced the ground water. If that first gallery, however, was the "heroic work" that a high official had once pronounced it,[87] another one below it, and necessarily longer, would have had the makings of a veritably Herculean enterprise. Perhaps in earlier days Antonio López would have been willing to take on such a task. But he was old, and in his final years preoccupied with ambitious endeavors closer to home—two adits being driven to cut through the Rich Hill of Potosí at a level lower than any earlier gallery. It is barely an exaggeration to say, then, that San Antonio del Nuevo Mundo, to which Antonio López had given renewed life in the 1670s through his most renowned underground exploit, declined and died with him in the closing years of the century.

SATISFACTIONS AND FRUSTRATIONS

Antonio López was one among what was by his day a broad and ancient stream of Spaniards who had come to America to make their fortunes. And he had been quicker and more successful in realizing that ambition than almost anybody since the wave of great conquerors in the early sixteenth century, a fact of which he was well aware and proud. But quite soon, it seems—certainly once the Amoladeras mines in Potosí began to reward him generously for his investment in them—wealth alone would not quite do. In this dissatisfaction, too, he was typical enough, for emi-

grants to the Indies in general wanted not only money, but higher standing, in both their own and others' eyes. It was a cliché of
the time that even surviving the trip to the colonies boosted a
man's self-esteem and bravado: all Spanish males arriving in Peru,
a viceroy had said in 1611, "place 'don' before their names and
call themselves adventurers (*soldados*). . . ."[88] For López, however, there may have been yet more to it than that. He was a younger son of a cadet branch of an old, but distinctly provincial, family.
The family, though noble and honorable, had never had any title,
but had lived in the shadow of the great Counts of Lemos. What
a coup it would be to transmute the silver wrested by effort and
investment from Potosí into honors that would elevate the name
of Quiroga to a standing close to that of those great Galician lords.
Whatever other worthy deeds various Quirogas had performed over
the centuries, none of them had ever managed to do that. It was
an aim that Antonio López pursued energetically—but unavail-
ingly—throughout the 1670s.

The first sign of López's pursuit of honors to match his wealth
appears early in 1669, when he issued a power of attorney to four
representatives in Madrid who evidently had access to the court and
the Council of the Indies. Describing himself in all his functions—
owner of mines and refineries in Potosí and its district, *mercader
de plata*, and *aviador* of silver producers of the *Ribera* of Potosí—
he empowered his agents to go before the king and the Council
to seek rewards for his services, of which he was sending written
evidence. At this point he had, apparently, nothing as specific as
a title in mind. Instead, any "offices or positions with honor and
income, lesser or greater, in these realms of the Indies or elsewhere, as his Majesty's will may be, either hereditary or for life"
would do.[89] At the same time, he and a nephew of his, don Benito
de Rivera y Quiroga, who had recently come to Potosí, sent a similar request for preferment to the viceroy of Peru, the Count of
Lemos.[90]

Over the following year or two, however, López's desire for honors grew in ambition as it also narrowed in focus. In a statement
of his services and merits submitted to Madrid at the beginning
of the 1670s, he announced for the first time what was to be his

enduring aspiration: the title of Count of Pilaya and Pazpaya. This was the name of the *corregimiento* in which his estate San Pedro Mártir lay, southeast of Potosí.[91]

It was partly, perhaps, the support of his powerful fellow Galician, the Viceroy Count of Lemos, that moved him to raise his sights so far and so fast. This is suggested by a memorandum from the Council of the Indies to the king in 1676, noting that "while [López] was consulting with the Count of Lemos about the grant of a title of Castile, by virtue of the powers that he held from your Majesty in that respect, the viceroy died without having brought the matter to any conclusion. . . ."[92] Since Lemos died in mid-1672, this consultation obviously took place before then. But more than viceregal encouragement may have lain behind López's evidently rising hopes of distinction, since it was in these years that the idea of selling titles of nobility in the colonies began to be put about in Spain as a way of raising money for government and defense. In October of 1672, in fact, an authorization was sent to the viceroy of Peru (presumably directed to Lemos, since news of his death could not have reached Spain so soon) to try to sell four titles of nobility to pay for the rebuilding of Panama after its sack in 1668.[93] And López was certainly to offer money for this purpose in a subsequent petition for the title he desired, which suggests that he was aware of the scheme to trade titles for cash.

There was undoubtedly, though, more to López's belief that he was worthy of noble rank than his capacity to hand over a sum of money to the Treasury. His petitions for a title constantly stress his past services to the monarchy, the great and growing scale of his mining enterprises, his present and future contributions to the common wealth through payments of taxes and wages, and, by no means least, the size and prosperity of his land holdings. Though wealth in the form of cash was highly desirable in a holder of a noble title, just as crucial for the maintenance of high nobility was the possession of wide estates. This old notion is clearly present in the mind of Antonio López de Quiroga, so innovative a man in many other respects.

The San Pedro estate had obviously been enlarged since López bought it in 1658. He proudly described it in the early seventies as occupying two river valleys which, combined, measured some fifty by forty miles (eighteen by fourteen leagues), and comprised grain fields, vineyards, and pasture for large and small stock. Many black slaves, well supplied with implements, worked this great area, which yielded products worth over 60,000 pesos a year. It was the land itself, though, and the numerous people on it that mattered, not so much the income—which paled beside the quantities of silver that López asserted his mines and refineries had produced, were producing, and would continue to produce. The aggregate output from his mines to date (1674) exceeded 40,000,000 pesos. He contributed annually to the treasury 100,000 pesos in royalties (70,000 in the three months October-December 1673 alone). No less than 400,000 pesos went every year in wages to the 900 Indians working in his various mills, adits, and mines, in payments for the 1,000 *quintales* of mercury that his refineries consumed every year (as much as was used by all the other *azogueros* in Potosí put together, he said), and in other mining expenses. Then, underlying these great sums, there was the silver producing plant itself: twelve mills (*cabezas de ingenio*) in Potosí and the district, and the various adits in the Rich Hill, Titiri, Ocurí, Laicacota, Porco, and San Antonio. And to round off this recitation of his wealth and worth, López threw in a passing reference to what must have been considerable urban property in Potosí: houses and other real estate bringing in 30,000 pesos a year in rent, and black slaves, silverware, and other valuable household effects in his possession. This all constituted, he thought, the "opulence and ample fortune" that were necessary for the "conservation, lustre, and ostentation" of such a title as Count of Pilaya and Pazpaya.[94]

Just as López's request for this title at the beginning of the 1670s demonstrates that his ambitions had grown considerably since he put in his general petition for any available office in 1669, so in the following three or four years did he raise his sights another sizable notch. Not having, apparently, heard anything from Spain after Lemos's death about his initial application for the title of Count, in 1674 he went on the offensive again, with a much

more detailed, explicit, and demanding statement of his requirements. He now requested full civil and criminal jurisdiction over
the 2,000 or so square miles of his proposed County of Pilaya and
Pazpaya, with rights to exercise capital punishment.[95] He wanted the right to appoint *corregidores* to administer the Province
of Pilaya and Pazpaya, and other officers of justice, without inter-
ference from the high court in La Plata. Then, because his estate
bordered on the lands of "barbaric and still unconquered Indians,"
he sought, in addition to the title of *Conde* of the territory he
owned, also that of *Marqués y Adelantado* of any of those wild
lands that he might bring under Spanish control; and he asked
for the right to appoint worthy men as *encomenderos*, or, roughly speaking, seignorial administrators, to govern the native people overcome in such conquests. The power to present candidates
for church posts in his domain was another of his demands. He
should be allowed to collect and keep for himself any tributes
payable to the crown by the Indians on his existing lands or those
to be taken; and since this would not be enough to offset his
expenses (the salary of the *corregidor* and priests, as well as the
cost of arms and powder), he should receive a supplement of 600
pesos a year from the Treasury. Finally, all his titles should be
hereditary in perpetuity, and the entire estate should be made
into an entail. In return for all these powers and concessions, he
now offered a donation to the crown of 40,000 or 50,000 pesos, to
be spent on rebuilding and fortifying Panama, though in his own
mind this was clearly no more than a token offering. What really
made him deserving of titles and authority was his previous contribution to the public good in creating wealth in mines and refineries. The millions of pesos he had drawn from the rock of the
Potosí district were a greater service to the monarchy than had
been rendered by any other subject.[96]

What lay behind this waxing of López's lordly ambition? Partly, of course, it may be that wanting more had become part of his
character, especially since, with few exceptions, he had been successful in getting more throughout his career in Potosí. Then again,
his many-pronged expansion in the late sixties and early seventies from the Rich Hill into mining ventures all over the district

may have given him a sense of widely broadening control and confidence. From being the master of mining in Potosí, he was rapidly becoming the master of mining throughout Charcas; and the full and ancient prerogatives of landed lordship seemed perhaps no more than was due to such a figure. At the same time, and most conveniently, the imperial administration in Spain decided to offer titles for sale. In 1675 the Council of the Indies (having authorized the sale of four titles in 1672) discussed a proposal that 150 titles might be sold across the Empire, on a sliding scale of prices: 25,000 pesos for the rank of Viscount, 35,000 for that of Count, and 45,000 for a Marquisate. The Council rejected the proposition on the grounds that although America contained people of sufficient quality to be worthy of such noble rank, those people were not the ones with the funds to lay out on such large purchase payments.[97] But, although the proposal was dismissed, some titles were still available for money. López evidently knew this, otherwise he would not have offered a gift of cash to the crown in 1674. Further, he was perhaps aware that the trend lay toward easier access to ever higher rank. The gift of 40,000 to 50,000 pesos that he offered to the crown in 1674 was in the range that had been suggested as appropriate for a marquisate. And it was for a marquisate that he petitioned in that year. The expansion of possibilities may, then, have helped expand his ambition.

Finally, the higher demands contained in López's 1674 petition may reflect a belief in him that he now had more powerful contacts at court than before, particularly in the shape of the Most Reverend *fray* Antonio de Somoza, Franciscan residing in Madrid, and Commissary General for his Order with responsibility for Peru. This office made Somoza in effect an adjunct member of the Council of the Indies, and it was to him above all that López entrusted the handling of his second bid for noble titles. Why should he have done so? The best answer is that Antonio de Somoza and Antonio López may have been distantly related. Somoza is a Galician surname, with origins in López's home territory, close to Monforte de Lemos;[98] and indeed López once wrote his full name as Antonio López de Quiroga y Somoza. Futhermore, it can hardly be a coincidence that the one and only recorded occasion on which he used this amplified form was in a report on his services sent to the king precisely in this year of 1674.[99]

It is tempting to suppose that when López discovered that Antonio de Somoza held a position in Madrid tantamount to membership of the Council of the Indies, he recalled a forgotten or neglected Somoza contribution to his ancestry, and added it to his name in a bid to elicit all possible support from this well-placed cleric. There is no record that López called on his Franciscan friends and allies in Potosí to send testimony of his worthiness to Somoza; but it would have been natural enough for him to milk his Franciscan connection for all it would yield.

Good contacts Antonio López may have had, and great were indeed his services. If he was not precisely the most valuable subject the king had ever had, as he apparently liked to think of himself, he was certainly worthy of being described, as the President of La Plata said in 1682, as the most *useful* vassal in Charcas, and one deserving of any favor.[100] But no title was ever to come his way. How, in a time when titles were openly for sale, can this be explained?

Although no record remains of deliberations in the Council of the Indies over López's application, it is not hard to guess what the objections might have been. López's terms were simply outlandishly demanding; and besides that, the peremptory tone in which they were expressed can only have made the councillors purse their lips in dismay. In the first place, López was asking for three titles for the price of one: Count, Marquis (which was higher), and *Adelantado*. Further, there had been no mention of selling the last of these. *Adelantado* was, in reality, a title that was no longer current. In Spain, it belonged to medieval times; in Spanish America, it had been granted during the period of conquest, but phased out over the second half of the sixteenth century. The title was, true enough, appropriate for López in his proposed role as pacifier and incorporater of wild Indians and remote frontiers, for it implied military governorship of just that sort of place. In Spain it had typically been given to men pushing forward the boundaries of Christianity during the Reconquest of the Peninsula from the Moors, and in Spanish America to many an intrepid *conquistador* in the early decades of Spanish presence in the New World. But precisely because *Adelantados* carried so much independent military, civil, and judicial authority, the crown had stopped issuing the title long before the end of the sixteenth century and had refused to allow the sons of the original holders to inherit it.[101] Perhaps López thought that by becoming an *Adelantado* he would be emulating that dim and distant Quiroga whose sharpened stakes, according to legend, had kept the Moors from Galicia, but the Council of the Indies can only have seen in his petition an outdated but decidedly dangerous bid for large independent powers.

It was this last aspect of López's series of requests, indeed, that most probably disinclined the Council to grant him any title. For, put together, the various prerogatives that he claimed threatened to add up to greater personal power and distinction than any private citizen of Spanish America—save, perhaps, Hernán Cortés in his final status as Marquis of the Valley of Oaxaca—had ever had. (And that comparison is strained, for Cortés, even after he had lost his governmental offices in the final years of his life, can hardly be counted as a private citizen.) However meritorious López's contributions to public wealth had been (though quite possibly his own great personal wealth was a source of misgiving), the central governing body of the Empire could hardly have given away to him the powers that he wanted, because they were powers that traditionally belonged to the monarch and to his high ministers in the Indies.

A close examination of the issues at stake here would be tedious. But, in brief, by claiming civil and criminal jurisdiction on his lands, López was proposing an attack on the judicial powers of the high court of La Plata, to which that jurisdiction belonged by all law and precedent. Similarly, by requesting the right to appoint *corregidores* for Pilaya and Pazpaya, he threatened to encroach on the authority of the viceroys, who, as chief executives in the colonies, were the officials who usually appointed *corregidores*. Again, by petitioning for the right to present candidates for parish priesthoods on his lands he seemed to want to arrogate to himself some of what was considered among the most extraordinary power belonging to the crown, conceded to it by the popes: the right of ecclesiastical patronage, or the privilege, in effect, of selecting priests for church posts in the Spanish colonies.[102] Finally there was López's proposal that he be allowed to name *encomenderos*, or quasi-seignorial administrators of Indians, in the territories he might subjugate. This was a privilege that *Adelantados* had possessed, to be sure;[103] in fact, it was precisely the sort of privilege that had led to the crown's anxiety about their autonomous influence in the colonies. But a deeper objection than this to granting López any such capacity was undoubtedly that the *encomienda*, or system in which *encomenderos* governed Indians, had, with a few exceptions, been viewed with great official distaste as a method of administering Indians since before the middle of the sixteenth century.[104] One large reason for this was that it had given excessive power to individual settlers. Once again, then, López was proposing to resurrect on his own domains an antique institution that had been undermined, though only with considerable effort, by an imperial government in Spain keen to prevent power accumulating in the hands of colonists. López gave every appearance of wanting to reverse that process—of wanting to set himself up, in fact, as exactly the sort of independent, seignorial lord that the crown least wished to see established in America. So it is no wonder that his bid for titles met with a deafening silence in Spain. Possibly, if he had been content with the mere *name* of Count or Marquis, which certainly would have given him unique distinction in Charcas, without pressing for the substantive powers of high nobility, the answer from Madrid would have been different. But, it seems, for him it was all or nothing. And nothing is what he got.

To excessively ambitious demands as a reason for the failure of López's quest for a title can perhaps be added some feeling among Councillors of the Indies that he was not quite suitable in social standing for a title. Evidence for this is indirect, but strongly suggestive. Late in 1675, for instance, the Council announced that no fitting candidates—that is, men with the desirable combination of both wealth and nobility—had appeared in Peru for the four titles that the viceroy had been authorized to

sell three years before.[105] Now López's application had been firmly on the table for three years or more both in Lima and Madrid
by that time; and so, since he obviously was not lacking the purchase price of a title, he was presumably among those found short
on nobility. Why this might have been so is open to speculation.
Conceivably, his occupation as a miner, and particularly as a miner who also engaged in a wide variety of commerce, was thought
incompatible with nobility. In the Council's remarks in 1675 on
selling titles in Peru a slight sense emerges that nobility and commerce were seen as contradictory.[106] And while the Quirogas' hon-
or and nobility as a family seem to have been genuine enough,
the fact that Antonio López came from a junior branch may have
told against him. Against him, too, may have been his Galician
origin, since Galicians, even more than other provincials, were
often the objects of disdain in seventeenth-century Spain.[107] Some
perception that he was not quite up to the mark socially would
also be consonant with a persistent oddity in his life—the lack of
don before his name. Despite the undoubted tendency of Spanish men emigrating to America to award themselves this sign of
social superiority, despite López's great wealth and influence,
despite the standing of his family in Galicia and the fact that
almost all the other Quirogas appearing in the Villa Imperial in
the second half of the seventeenth century did use the *don*, no
document written in Potosí or elsewhere in America that mentions Antonio López applies this label of distinction to him. The
omission is remarkable, since the use of *don* became ever more
free and easy in colonial Spanish America as time wore on; and
if there were ever anyone in Potosí who might have been thought
to have earned a *don* by gaining local eminence, no matter what
the drawbacks in his background, that person was certainly Antonio López de Quiroga. But don Antonio he never was,[108] any
more than he was Count, Marquis, or *Adelantado*.

After the mid-1670s, López's pursuit of noble rank became less
energetic. He restated his request, though without much force,
in a letter to the king written from Potosí late in 1679. After that,
there is silence on the topic until 1690, when, as if going back in
resignation to his original and, for him, humble petition of 1669,

he empowered a new set of agents in Madrid to seek for him unspecified "grants of offices and honorable positions" from the king and the Council.[109] This new effort produced no result, even though by this date titles had been sold in the Indies,[110] and indeed a Mexican silver producer, don Juan Bravo de Medrano, from Zacatecas, was successfully negotiating the purchase of the title of Count of Santa Rosa. Being a miner, then, was not an insuperable barrier to gaining an aristocratic title. Bravo's success suggests strongly that it was other obstacles, such as those discussed above, that stood in López's way.[111]

It is not true to say, of course, that López was totally without a distinguishing "label" to his name, because since the late 1650s he had held the honorary military rank of Captain, and in 1672 he was "promoted," as it were, to the next level of *maestre de campo*, or colonel.[112] This seems likely to have been the work of Viceroy Lemos, both because of his connection with López and because the viceroy was the supreme military officer in his jurisdiction. The promotion implied no great distinction, though. López's father-in-law, Lorenzo de Bóveda, had become a *maestre de campo* in his late years, and he, although a wealthy merchant and solid citizen, was certainly a less remarkable figure than Antonio López had become by 1672. Besides, in late seventeenth century Potosí, there were at least a half dozen *maestres de campo* besides Antonio López, including one of his sons-in-law, don Juan de Velasco, and two of his brothers-in-law, don Joseph, and don Juan Antonio, de Bóveda y Saravia.

No military duties seem to have gone with these honorary ranks, at least in the normal course of events. On the other hand, holding them may have given the incumbents a prominent and gratifying ceremonial role to play in municipal affairs, with some military overtones. Arzáns's account of the part that López took in celebrations held in 1674 to mark the canonization of Saint Francis of Borgia conveys the spirit of this, although a zero should perhaps be docked from his flamboyant numbers.

The fiestas were crowned with a magnificent procession, in which the *maestre de campo* Antonio López de Quiroga showed his greatness. He invited 330 nobles to take part, distributing to them arms, powder, silk stockings, splendid beaver hats, and brightly-colored ribbons, all with much liberality. He himself wore so rich a costume of dark brown cloth, embroidered with gold thread and studded with pearls and precious stones, that it was thought to have cost 40,000 pesos, not to mention the many resplendent jewels in his hat, among which was a diamond of rare size costing 6,000 *escudos* in Spain, whence he had caused it to be brought.[113]

There is no account besides Arzáns's of this event, but López's described part in it seems at least to match his pretensions at the time. A mere honorary *maestre de campo* he might be, but his wealth enabled him to put himself at the head of a mass of armed (if only ceremonially armed) men and lead them through the town as if he were truly the high lord he hankered to be. Alas for López, Arzáns's "nobles" can have been so only in the most general sense— the leading lights of Spanish society in Potosí rather than men of long and honorable Spanish lineage.

The rank of *maestre de campo*, with its militaristic ring, seems aptly suited, also, to another of Antonio López's great expansionary projects of the 1670s, one in which many of the propelling forces in his career and personality—ambition for wealth, ambition for fame and honor, emulation of the heroics of *conquistadores*, but all this entwined curiously with a pervading sense of the power of financial investment—are visible in combined play. This was the exploration, conquest (if necessary), evangelization, and exploitation of the fabled realm of the Gran Paitití.

Almost from the beginning in the 1530s, when Francisco Pizarro and his men had revealed and seized the wealth of the Inca empire among the high ranges of Peru, Spaniards had gone off, some striding, others trudging, in search of similar riches on the inland plains east of the mountains. In the early days, many went because they had arrived too late to claim a share of Inca booty. Many others went later because it seemed that the lowlands, vast, forested, and mysterious, *must* hold hidden kingdoms and opulent rulers. Many never came back. And those who did never had anything concrete to report. But this, if anything, merely intensified beliefs and hopes. And so, in the north of the continent, the myth of El Dorado grew ever more alluring. And farther south, in the region where Bolivia, Peru, and Brazil now meet, the Great Paitití became a figure to capture the imagination.

Behind the myth there may have lain a modicum of Incaic reality. Spanish chroniclers of the sixteenth century relate that during the expansion of the Inca empire in the mid and late 1400s, the rulers in Cuzco sent out men into the lands of the lowland tribes to the east; and that these men penetrated as far as the region where repute was later to place the Paitití, building two fortresses there. The place was not, in reality, forested plain, but low, bare hills rising out of it, now called the Serra dos Parecis, and lying inside the arc formed by the Rivers Guaporé, Mamoré, and Madeira [114] (See Fig. 5). Small numbers of warriors were reportedly despatched to ensure the compliance of local tribes, but no further Incaic settlement followed (as would be expected, since the highlanders had no love for the forests) until the Spaniards entered the Inca heartland in the 1530s. Then, by one chronicler's report, 20,000 people fled from the Inca center to take refuge among the hills beyond the jungle barrier. Among them were craftsmen, including silversmiths.[115]

It was this reputed migration that underlay the myth of the Great Paitití, creating among seekers of fame and fortune the hope of a lost Inca state that would be bigger and richer than the quite real remnant of the Empire maintained until 1572 by the Inca Manco and his successors at Vilcabamba.[116] One of the early Spanish expeditions that went in this direction, despatched by Pizarro himself in 1538, had as its *maestre de campo* (in this case, far from an honorary appointment) the Rodrigo López de Quiroga who later became Governor of Chile;[117] though if Antonio López knew this, he unaccountably left it out of the telling of his own efforts to pin down the Paitití. Two-thirds or more of the men on this journey perished, and the survivors had nothing to report besides forests and primitive tribes. But the legend grew. In 1564 the high court of La Plata told Philip II: "A hundred leagues, at the most, from this city a very rich land has reputedly been found where there are many Indians and much gold and silver, so that, even if

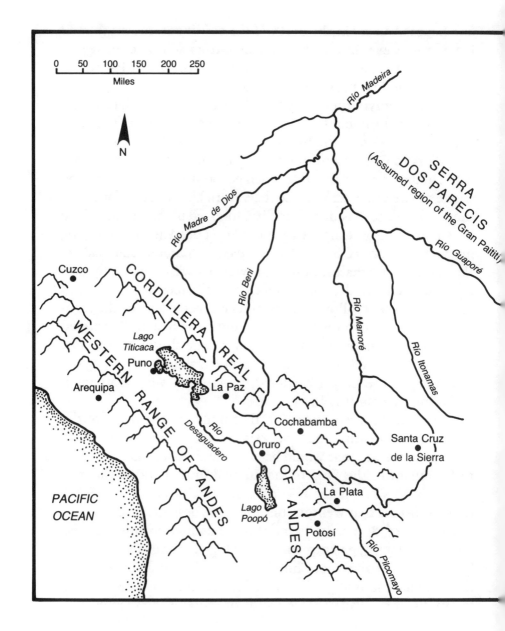

Towns of Charcas, and the Believed Domain of the Gran Paitití

it is only a tenth as wealthy as people say, it will be better than Peru."[118] And the refrain was picked up a few years later by no less than Viceroy Toledo, who, hard-headed as he was, was moved to even greater heights of enthusiasm by reports from explorers of the plains: "In the province of the Paitití are mines of silver and gold, and a great quantity of amber. The news is of the greatest amount [of these] and the greatest wealth yet found in all America."[119]

In these sixteenth century reports it is difficult to tell whether *Paitití* means a place or a person, but by the early seventeenth century the notion of a great ruler has become firmly established. An account of 1623 speaks of an "infinite people" living on islands in a lake formed by the rivers of the plain, "and the lord of them all is called the Great Paitití." A "province of women who live without men" lay nearby.[120] Amazons are a new embellishment of the story, and one that did not persist. But what did last was the conviction that the subjects of the Paitití were descendants of Incas, and that much wealth in precious metals could be expected among them. These notions were perhaps given a tinge of spurious officialdom when in 1636 the President of the court of La Plata, considering whether to authorize yet another expedition to seek the Paitití, gathered no fewer than ten earlier reports that sounded these themes clearly. At the same time, a tendency to run together the Paitití and El Dorado appears, a confusion that on the one hand reflected the wild imprecision of geographical knowledge of the interior that still prevailed, while on the other adding further allure to tales of this supposed neo-Inca land.[121]

The idea of organizing an attempt to explore the approaches to the realm of the Paitití, and then the realm itself, if it could be found, was planted in Antonio López de Quiroga by a Dominican friar who during the 1660s had spent two or three years on the plains carrying Christianity to their inhabitants. This was one Father Francisco del Rosario, a Galician, like so many other people with whom López had important dealings in Potosí. Another Galician it was, the Viceroy Count of Lemos, who gave official sanction to the attempt; and yet a third, López's nephew don Benito de Rivera y Quiroga, who led it.

Father Rosario was in the group from Potosí that went late in 1668 to meet the viceroy at Laicacota. The outcome of the discussions that he and López had there with Lemos was governmental approval of López's plan to finance an attempt to "conquer and pacify" the native peoples known as the Mojos, Chunchos, and Raches, who lived on the plains between the mountains and the territory of the Paitití. The immediate aim was to convert these people, for although, Rivera and Quiroga said, Father Rosario's earlier efforts had gathered "a good harvest for Heaven, he realized the necessity of the sword to defend what the Gospel taught, because the Indians easily changed their religion. . . ."[122] But beyond this conversion lay the greater lure of the rich Paitití itself, of which the viceroy now gave Rivera y Quiroga the title of Governor, in anticipation of his success in both finding and overcoming it and its ruler.

López and nephew went back to Potosí and lost no time in putting together a reconnaissance force. On March 6, 1559 an earlier holder of rights to conquer "the infidel Indians of the Mojos and Raches [provinces] in the empire of the Great Paitití," Captain Pedro Fernández de Salamanca, ceded them to Rivera y Quiroga,[123] who, however, attracted to his company one of Fernández's experienced lieutenants, don Juan de Mesa y Zúñiga, making him his *maestre de campo*. Then Antonio López bought thirty black slaves, paying the tremendous sum of 1,200 pesos each for them, he said, because he wanted the strongest available, who could withstand the heavy labor of hewing a way through the dense forest ahead of the main Spanish force. "Cutting down very stout trees that are more than 140 feet high," he thought, was more than Spaniards and Indians could manage. Then there had to be tools for the job—hammers, crowbars, machetes, and axes.[124]

With an expeditionary force consisting of these black slaves, "a few soldiers and many pioneers [*gastadores*]," and don Juan de Mesa and Father Rosario as guides, Rivera y Quiroga went off in 1669, probably before the middle of the year. The group advanced twenty-six leagues (seventy to eighty miles) with great effort, he reported the following year, crossing "rugged cordilleras and thickly forested hills" until it reached the plains. (No route is mentioned, but the sequence of mountain and forest fits a crossing of the eastern Andes northeast of Cochabamba and a descent through the densely wooded and humid Yungas zone beyond them). Up to this point, it was apparently possible for the Spaniards to ride (and Antonio López even refers to the use of litters);[125] but now advance was on foot. And there, on the edge of the plains, Rivera reported, "In a stone fortress made by the Inca kings I set up a chapel that was dedicated to the Virgin of the Rosary, patroness of this conquest." The sign of Incaic presence was doubtless encouraging, as was the abundance of people in the land. But precisely because of this populousness, a larger force now seemed necessary, and a council of war agreed that the present group would go no farther. Rivera y Quiroga was back in Potosí by March 1670 to raise more troops, but eager to be away again once the rainy season ended (as it does soon after then).[126] Nephew and uncle proudly announced to the Council of the Indies that they had opened a route, at great effort and expense, to the beginning of the plains of "that New World" in which the empire of the Great Paitití lay.[127]

Arzáns records that López and Rivera went about raising men for the new and larger force with some elaborate fanfare. A royal standard for the expedition was publicly blessed in Potosí, and on that same day, "two companies of two hundred men each were asembled, with the first under the command of the Governor don Benito, and the other of the *maestre de campo* Antonio López de Quiroga. Both companies paraded with a great show of dress and jewels, and, arriving in the plaza, formed into battle order, and then attacked each other with much skilled play of sword and harquebus, though without injury."[128] Some of this may be Arzáns's typical embroidery of events; on the other hand, extravaganzas like this had often been performed at the beginning of conquests, explorations, and what the Spanish generally called "entries" (*entradas*) into new territories, from earliest colonial times. Furthermore, López can easily be imagined revelling in this sort of martial play-acting, since it would have allowed him some small degree of physical participation in the heroic enterprise that he was organizing and funding, but which age and business duties prevented him from engaging in any more closely.

The mock battle attracted recruits, apparently, though it did not raise enthusiasm to a pitch where men would go off at their own expense. López was later careful, as usual, to point out how much it had cost him to raise and equip this band of explorers. Besides the 36,000 pesos that he had already paid for his thirty robust slaves, he now laid out 120,000 more in wages and supplies for eighty-seven armed men (*soldados*) and officers to command them. He also refers to "herds of livestock of all sorts" that went along with the expedition. At the very least this suggests that besides horses for riding, and mules or llamas for freight, the party may have been supplied with pigs—a form of mobile food supply that had sustained Spanish explorers and conquerors of America since earliest times.[129]

Now López prided himself, and many remarked, on how completely his estates, mines, and refineries were equipped, and he evidently undertook the job of outfitting his exploratory force with the same care and forethought. But perhaps he went too far, for the expedition came to an abrupt halt when a majority of its men made off with a year's worth of supplies. This happened after the party had retraced the earlier route through the Cochabamba valleys, and over the eastern ranges into the Yungas and eventually the plains. There they had made contact with Indian villages, which, according to López, immediately gave themselves in allegiance to the crown. At this point, Rivera y Quiroga decided he had to go back to Cochabamba to attend to "some necessary affairs."[130] During his absence, the main body made camp, while a reconnaissance group of eighteen men, led by Juan de Mesa, went ahead down river.[131] But when this group returned, they were told by local Indians that all the rest had fled, taking all the baggage. Rivera y Quiroga tried to round up the fugitives, but, getting no cooperation from Spanish officials, including the court at La Plata, abandoned the effort, and, with a few faithful men, returned to the forests and located four more Indian villages, which he also brought into allegiance to the monarchy. Then, in 1672, he built eight of what López calls "very capacious hostels (*ventas*) where people can be lodged and supplies stored," spaced ten to twelve miles apart, and brought in donkeys and oxen for freighting. The land, so López told the king in 1674, was healthy and peaceful. There were wild honey, nutmeg, and cinnamon to be had. Only one Spaniard had died there in the three years since exploration had begun in earnest—and that was of old age.

It was, evidently, exceedingly irritating to López that his well-laid plans had gone so far awry, and reality fallen so far short of what he had aimed for: wild honey in place of gold and silver; forest tribes in place of the legendary polity of the Gran Paitití. While during the 1670s he makes no written criticism of his nephew's behavior, he can only have wondered whether things would not have gone much better if Rivera y Quiroga had not abandoned his command at precisely the point where it most needed direction and exhortation. But for the moment he vented his anger in a general blast. "The people of this [land of] Peru are lazy and given to vice, enemies of the labor that so heroic a conquest requires."[132]

This generalization might, however, have its exceptions, for López had heard that "in Santa Cruz de la Sierra there were good people for conquering mountains,"[133] and so he asked Viceroy Lemos to confer on Rivera y Quiroga the office of Governor of Santa Cruz, so that he could make effective use of this promising human resource. (The absence of mountains between Santa Cruz and the supposed location of the Gran Paitití might seem to have made the *cruceños'* special abilities less useful to López than he thought, but he probably knew no more than most others at the time about the geography of what is now eastern Bolivia.) Lemos obliged, at least to the extent of granting an interim title. And from that base in Santa Cruz, both physical and official, Rivera y Quiroga continued to lead small expeditions to the north throughout the rest of the seventies. López thought that settlement of the region north of Santa Cruz would make Cochabamba the greatest town in Charcas, partly because "the traditions that there are gold and silver there are very strong," and also because "according to Franciscan chronicles there are more than 1,500 leagues [of territory] to conquer, all peopled by unbelievers; and once this land is subdued, it is said that one could go to Spain and return in six months."[134] López's notions of geography might leave much to be desired; but, in this conception of a route to Europe from the eastern Andes down the rivers to the sea, he could call on a long, if undefined, tradition of Amazonian exploration by the Spanish and Portuguese. It was a dream, however, that from Bolivia, at least, still today remains to be realized.

The prospect of a quicker route to the Atlantic and Spain, however, was workaday stuff when set beside the earlier vision of a mysterious high culture pursuing its secret ways in the center of the continent. López's retreat, as early as 1674, to this practical benefit of his explorations perhaps indicates that after only a few years of supporting Rivera y Quiroga's efforts, discouragement, or more likely, the scepticism of the tough-minded businessman, was working to dispel this particular heroic dream. For whatever reason, nevertheless—probably a mixture of the practical and the visionary—he continued to back Rivera. By the end of the seventies nephew and uncle claimed to have spent more than 350,000 pesos on explorations. And, so Rivera pronounced in 1679, as a result of this expenditure, and the running of "excessive risks, I managed to plant the royal banners of your Majesty in places where no one has been able to go before, . . . subjecting two provinces to the obedience of your Majesty, whence plentiful fruit for Heaven has been plucked."[135] Sixteen previous *conquistadores*, beginning with Pizarro, had attempted to find the *Gran Paititi y Dorado*, he continued, now combining the two myths, perhaps to add merit to his efforts. None had succeeded; and neither had he, of course, although with "inexpressible hardships and efforts" throughout the seventies he had gone deeper into "the gulf of the cordillera that surrounds these provinces from Portobelo to Pernambuco" than anyone before him. Behind all his efforts, Rivera reminded the king, stood Antonio López de Quiroga, inspired by his zeal to serve his God and his monarch. It was López who had "ordered" him to essay this conquest—the same López, he continued for the king, who was already so great a servant of the crown by virtue of his payments for royalties and mercury, and his generous donations to the king, that no parallel to him could be found in modern or ancient history. But this had not seemed enough to the *maestre de campo*, who had therefore resolved to serve the king in yet new and different ventures.[136]

By the time Rivera y Quiroga wrote this, however, Antonio López had already decided at least to reduce, or possibly to end, his backing for his nephew's expeditions. The practical man had taken command again by 1680. His mining expenses, especially those connected with driving adits, were high (and he was still embroiled in efforts to recover from his partner in the adit at San Antonio, Espinosa Patiño de Velasco, half the costs of that undertaking). He thought that the king might now reasonably be expected to take over some of the expenses of exploration in the eastern plains; or, as Rivera y Quiroga put it in his usual mellifluous tones: "It seems that Heaven does not desire the complete fulfillment [of our purpose] unless your Majesty contributes some whit of your royal power as our full partner. . . ."[137]

The Council of the Indies asked the high court of La Plata to comment on this request,[138] and when it did so, the answer was at best cool. The court conceded that López had spent heavily on explorations—not perhaps the 350,000 pesos that he alleged, but probably over 200,000. But there was no sign that spending this great sum had yielded any useful result at all, besides a little evangelization of the forest Indians, which was already under way in any case. No harm had come from López's and Rivera's attempts, and they deserved some royal reward for their effort and expense.[139] But on the wisdom of royal aid for further efforts directed toward the Paititi, the court's letter is quite silent. And evidently this was taken in Madrid as a judgment on Rivera's request, for no royal funds were ever forthcoming to support new forays by him into the eastern forests.

Nor, after 1680, did Antonio López contribute anything more to his nephew's ventures in the east, which continued vigorously, at least according to the nephew himself. Gradually acrimony between Rivera y Quiroga and his uncle grew up over this lapse in backing. In 1696, Rivera, still styling himself Governor and Captain General of the Great Paititi, declared that he had worked incessantly since the late 1670s to bring the people of the plains into the fold, but was now "totally unable to continue and to give aid to the more than 300,000 souls who have rendered obedience to his Majesty" in his domains. How he arrived at this immense figure he neglects to say. He petitioned the court at La Plata to order López to resume the subsidies that the two of them had agreed on long ago.[140] When asked by the court to comment on this, López simply stated in October 1696:

What I have to reply is that in the conquest of the Gran Paitití, with the desire that the Indians who inhabit it should come into the fold of our Holy Mother Church, I have laid out more than 300,000 pesos; and that [my nephew], with the slaves that he took from Potosí, created an estate that he still possesses; and that the better to furnish supplies [for the conquest] I bought the estate of Biloma, which is in the province of Cochabamba, which my nephew later gave in dowry to his son-in-law General don Diego de Quiroga; and that today, being very old, I am not in a condition to be able to continue in any matter connected with the said conquest. . . ."[141]

And there, it seems the matter still stood when López died a little more than two years later. Whether Rivera y Quiroga had really taken advantage of López's largesse to the degree that his uncle implied is not to be known. It is suggestive, though, that Arzáns, writing about the Paitití venture only a few years later, commented that in the province of Mojos (or more plausibly on the edge of it, in the Yungas north of Cochabamba) Rivera had tried to establish a sugar cane plantation, and other sorts of farms and gardens, "and so passes his life most agreeably, without any further desire to go off in quest of the Paitití."[142]

It is not quite accurate to imply that López only gave for the enterprise of the Paitití and gained nothing in return. For in 1686, by one report, he had received from the king some recognition of his efforts in that direction—but only in the form of the right to use, presumably on any coat of arms he might have, the motto *Non plus ultra*.[143] It was not much, especially for a man who wanted high titles, but it may have pleased him on two counts. First, the obvious allusion to the positive form (*Plus ultra*) of the motto, adopted long since by the Spanish Hapsburgs, could only have been flattering. And the imperative that must be the first sense of the phrase—No farther!—though possibly irksome to one more used to giving commands than to receiving them, may really have been welcome confirmation of a decision that he had already taken to abandon distant fancies, and stick to the concrete tasks that he knew and could direct so well.[144]

FAMILY MATTERS

If expansion is the prevailing theme of López's activities in the 1670s—physical expansion to new mining centers in the district, psychological expansion to new aspirations for social rank and historical renown—then, in a small but significant way, his family affairs in the decade also take this direction. For this was the time when his daughters were reaching womanhood, and when the first of them married, establishing new connections for López that he was quick to exploit.

Doña Felipa Bóveda de Saravia and Antonio López de Quiroga had only two children, both of them girls. The birthdates of doña Lorenza (presumably so named in memory of her grandfather, Lorenzo de Bóveda) and her sister doña María, are not clear. But both were alive in 1664, when their father registered two mines in the Amoladeras deposits in their names.[145] Both were described as still being minors in 1675[146]—as indeed they almost certainly would have been, since the legal age of majority was twenty-five, and their parents had married in the early fifties.

In the following year doña Lorenza herself was married. Her husband was don Juan de Velasco, a man born in Spain, at Tordelaguna in La Mancha, and probably rather older than she was, since before coming to Potosí he had been *corregidor* of Conchucos, in the northcentral mountains of Peru, an office to which he was appointed in 1671.[147] He was also a secretary, or scribe, of the Tribunal of the Inquisition in Lima, the central institution of the Holy Office in Peru. Whether his move to Potosí was in some way connected with this position, or whether in fact he exercised it there at all, is not known. But even if he kept no more

than the form of the office, in some honorary way, his associa-
tion with the Inquisition very likely gave him considerable standing and, even more, power of intimidation in the Villa Imperial.[148]
To top it off, don Juan was also a *maestre de campo* of the honorary sort, like his newly-acquired father-in-law.[149]

The marriage took place late in October 1676. Antonio López
despatched his daughter into her new estate with a dowry of
100,000 pesos, to which the bridegroom added a gift of 20,000
pesos, "on account of the honor, virginity, and chastity" of his
bride.[150] The most valuable single object in the dowry list was a
four-poster bed with a frame of gilded *granadillo* (a hard, dark-
red wood), decorated with bronze finials, and complete with a
spread and skirt of crimson damask, embroidered with Milan gold.
This was assessed at 6,000 pesos. Next in value were six black
slaves (four men and two women), at 800 pesos each. Eight tap-
estries, showing historical scenes, were appraised at a total of 4,000
pesos. Two items were listed at 1,000 pesos each: a writing desk
of ebony and ivory, with corner-plates and feet of gilded bronze,
and a sedan chair with six glass panes, upholstery of crimson vel-
vet, and gilded nails. A dozen paintings illustrating the months
of the year, in gilded frames, and "painted in Spain," were adjudged
to be worth 100 pesos each. Other useful or decorative goods giv-
en in dowry included two dozen elegantly upholstered chairs, two
oriental carpets (one sixteen feet long, and the other fourteen), a
pearl necklace and choker, emeralds, diamonds, objects of gold
and crystal, 7,000 pesos' worth of silverware (plates, pitchers, salt
cellars, basins, and cutlery), and 6,000 pesos' worth of fabric goods,
such as table cloths, dressing gowns, chemises, handkerchiefs,
bedclothes, and pillows. The value of all this came to 50,925 pesos
and 6 reales, according to the two appraisers appointed by Velasco
and López de Quiroga. The balance to 100,000 was made up in
silver coin, in twenty sacks.[151] This may have been the only part
of the dowry made in America, for with the possible exception of
the silverware and the sedan chair, everything else is very likely
to have been imported. Local products were simply not thought
fine enough for the occasion. And if what was in the dowry is
any sign of what López had in his own house, it is probable enough

that the *maestre de campo* himself lived amidst considerable luxury of furnishings.

Whether doña Lorenza's mother attended her wedding is uncertain, because a reference to her in 1676 suggests that she may already have died by then,[152] and she was certainly no longer alive by August of 1677.[153] She was, by Arzáns's account, a model of good sense, serenity, and beauty, with whom López had lived in exemplary accord and chasteness.[154] Given Arzáns's tendency to eulogize the moral, this judgment should perhaps be taken with a little salt. But neither is there evidence to suggest anything to the contrary. Doña Felipa was buried in the Franciscan church of Potosí, in a tomb that Antonio López was later to share.[155]

In Velasco, López had gained a relative of standing, authority, and administrative ability (as *corregidor* of Conchucos, don Juan had "given a very good account of himself," according to official judgment).[156] He soon put his new son-in-law to use, sending Velasco off south in September 1677 with a sweeping power-of-attorney to collect debts and take charge of his mining and refining operations in San Antonio del Nuevo Mundo.[157] Velasco, nevertheless, may have preferred in the long run to remain his own man. For there is no further evidence of his acting as an agent or manager for Antonio López after the 1670s. By 1681, in fact, he had taken up producing silver in his own right, and was reported as operating four stamp mills in that year, a number well above the average for *azogueros* in the Potosí district. And so he continued, apparently, until his death some fifteen years later, in 1697, adding in the meantime to his honors and titles a knighthood in the military order of Santiago.[158]

If Antonio López, however, seems to have lost the services of one able son-in-law at the beginning of the eighties, he gained the help of another, equally competent and perhaps more energetic, not many years later, when his other daughter, doña María, married a man active in commerce and local government in Potosí, Miguel de Gambarte.[159] The first traces of Gambarte's activities in the town date from the mid-1670s, and in 1676 he is described first as a resident, and then, a few months later, as a householder in the Villa Imperial, and an honorary Captain as well.[160] He came to Potosí from the south, having made his first landfall in South America at Buenos Aires in 1669 as a speculative merchant. With two partners, he had outfitted three ships in Spain and filled them with Spanish cloth goods for sale in the Río de la Plata. Obtaining permission from the crown to take this cargo to what was a restricted port of entry had cost him 28,500 pesos. And when he got there, he found the market in and around Buenos Aires unable to absorb what he had brought (partly, he said, because illegal Portuguese traders had recently cleaned out all the cash reserves of the inhabitants). So, having decided with his partners that they should take the ships back to Spain with a cargo of hides, he undertook to stay on in the Plata region until he found an outlet for the cloth. Eventually, in 1674 or 75, he managed to dispose of some of it northward, to Paraguay, in exchange for *yerba mate*, with the idea of sending this bitter tea up to Potosí for sale, since that was the market, in his view, where there was most demand for it.[161]

Such, at any rate, is Gambarte's own account of how he came to enter the commerce of Potosí, as an innocent tea merchant. And indeed his *mate* does seem to have reached the Villa Imperial. But the authorities believed that not all the goods that he had imported from Spain had been sent to Paraguay. Many of them, they suspected, had been spirited to Potosí along with the *mate* leaves; and that was another matter altogether, since goods imported under special permit into Buenos Aires could not be legally sold in Charcas. So, hardly had he arrived in Potosí when Gambarte found himself under prosecution as a smuggler.[162] And, despite his protestations, in the end the Council of the Indies found convincing enough evidence that he had brought goods up illegally from Buenos Aires to sentence him to a fine of 2,000 pesos and four years' exile to the outposts, or *presidios*, that guarded the frontier between Spanish settlement and hostile Araucanian Indians in southern Chile. Gambarte bought his way out of this punishment, however,[163] and struck out on a rapidly rising career in Potosí. A sure sign of his almost instant acceptance there is his selection by the *cabildo* in 1679 as one of the two *alcaldes ordinarios*, or first-instance magistrates, for that year.[164] This was not a position that went to nonentities or insubstantial men. He again held the office in 1684.[165] All the while, his trading forged ahead. Whatever trouble *yerba mate* might previously have landed him in, it still attracted him as a trading commodity. In 1678, he can be found sending a consignment of it to Chucuito, on Lake Titicaca.[166] And in 1681, his remittance of a large sum of money to one of his business partners of 1669, in fact the owner, now back in Seville, of the vessel in which he had himself travelled to Buenos Aires, suggests strongly that he was still active in transatlantic trade then, and quite possibly an importer of European goods into Potosí.[167]

The first record of any business arrangement between Gambarte and Antonio López dates from 1687, by which time the two were probably connected through Gambarte's marriage to doña María.[168] It is not difficult to imagine López's being attracted by the slightly swashbuckling quality that Gambarte's South American career up to this point seems to display. From now on they were to be close business partners, with Gambarte perhaps taking an ever greater role as López passed through the final years of his life. Gambarte outlived him by several years, dying in Potosí on June 8, 1706—by which time he was, like his long dead brother-in-law, don Juan de Velasco, a knight of the Order of Santiago.[169] His earlier smuggling transgressions had, apparently, been forgotten or forgiven. As López de Quiroga's son-in-law, and a venturesome businessman besides, Miguel de Gambarte was an obvious choice of partner for the *maestre de campo* to make. But Gambarte was only one among many associates and managers whom Antonio López selected from among his relatives, creating with their collaboration a network of contacts and power that in the end covered most, if not all, the places in Charcas where he held substantial economic interests. This structure began to come together in the 1670s, as can be seen from the roles López assigned to his nephew, don Benito de Rivera y Quiroga, in the exploration of the Paitití, and to his first son-in-law, Velasco, in the Lipes mining zone. But the full development of this web of influence and representatives belonged to the eighties. Possibly by then Antonio López was beginnning to feel the effects of advancing

age, and preferred increasingly to delegate reponsibility to trustworthy family members. But this development in his business
method, and broadly speaking in his career, may well also have
grown out of the disappointments of the seventies. He had made
his bid for public recognition and display by pursuing titles of
nobility, and, in a different way, by assuming the role of *conquistador*, at least at one remove, of the Gran Paitití. When both these
schemes came to naught, he may perhaps have resigned himself
to accepting a less conspicuous position in the Spanish world than
he would have liked to hold. But, at the same time, it is as if he
also resolved to fortify his economic achievement, which no one
knowledgeable of such things could deny, by weaving for himself a tissue of covert influence and power in Charcas with the
help and participation of family associates in whom he could place
great trust. To the great expansion of Antonio López's activities
in the 1670s, then, there succeeded a consolidation of his interests and activities, with himself, spiderlike, at its controlling center. And this consolidation is the final stage and culmination of
his career.

4

AUTHORITY

IN GENERAL IN SPANISH AMERICAN COLONIAL TOWNS, RICH MEN liked to hold local office. Some looked for posts that gave power or at least the opportunity to gather further wealth, such as the municipal magistracies or the position of town constable. Others preferred functions of a more honorary sort, such as acting as royal ensign in a community or becoming an alderman. Some of these posts could be bought outright; others, such as the annual magistracies, or *alcaldías ordinarias*, were assigned by election within the town council itself, although the influence of money could never be far removed even from elections. Potosí was no exception to these generalities. There from the beginning, *azogueros* had generally dominated membership of the council, and they had been the ones inclined to put out the heavy payments needed to secure other desirable offices.

Antonio López, despite being in all likelihood the one man in seventeenth century Potosí who could most easily afford to acquire some influential or high-sounding post in the Villa Imperial, nonetheless never did so. It is hard to say whether this was a matter of choice or circumstance, though the signs point perhaps more to the second than the first. Give López's obvious love of public display, an aldermanship, or *veinticuatría*, should certainly have seemed attractive to him. But nothing indicates that he ever tried to buy one, which perhaps suggests that he felt that for some reason the town council would not be receptive to him. More definite evidence in this line is the fact that when he *was* considered for elected office by the council, he fared dismally. In 1676 and again in 1678 he was among the thirty or so men considered for the year's two *alcaldías ordinarias*. Whether he or someone else put in his name on these occasions is not recorded in the minutes. In any event, in 1676 he received only two votes from the assembled *veinticuatros*, and in 1678, only one; while the two winners in 1678 took eighteen and fifteen votes respectively.[1] So he clearly had minimal support in the council for that position, and possibly he could not have expected better backing for any other post.

Why this should have been so can only be speculated upon. An obvious possibility is fear among the council members that the presence of such a rich and determined figure in their midst would simply overpower and overshadow their own lesser selves. Another reasonable apprehension would have been that to give López judicial authority as a magistrate would make him intolerably powerful in Potosí. But again, also, a question must arise about López's respectability. Is there not, perhaps, some common thread running through this absence from municipal office in the Villa Imperial, the lack of "don" before his name, and the failure of his attempts to win a title of nobility from the crown? Something, it must be suspected, was found to be not entirely proper about the *maestre de campo* Antonio López de Quiroga.

EXTENDED FAMILY AND INFLUENCE

But if Antonio López did not, for whatever reason, occupy any public office, then quite the reverse is true of his relatives. He, in 1678, was very far from being elected an *alcalde ordinario* of Potosí, but in the following year his future son-in-law, Miguel de Gambarte, held that job, and did so again in 1684. And others from López's close family circle had already taken up appointments outside Potosí. The first to do so was his brother-in-law don Juan Antonio Bóveda y Saravia, who was appointed *corregidor* of the province of Pilaya and Pazpaya by Viceroy Lemos in 1671.[2] This, of course, was precisely the region in which Antonio López's country estates lay. It was convenient for him to have a close relative as the prevailing authority there, especially because, a *corregidor* had at least some of the judicial powers that López was about to seek as Count of Pilaya and Pazpaya. And it would be no surprise if López had used his friendly connection with Lemos to have his brother-in-law named for that particular job. Don Juan Antonio went on to a career that mixed mining with further public office, all with close ties to López. In 1677 he bought mines in a part of the Rich Hill of Potosí called "Laca Socavón," where Lopez was also then operating.[3] By that time he had acquired the title of *maestre de campo*, which he was later to raise to the higher honorary rank of *general*.[4] By 1681 Bóveda y Saravia had become a *corregidor* again, this time of the province of Chayanta, where, of course, Antonio López had major mining and refining interests in Ocurí, Titiri, and Aullagas. Bóveda was still mining too, as his purchase of mines at Ocurí in that year shows.[5] For a royal administrator to engage in any sort of business within his jurisdiction was, naturally, against the law,[6] though by this time in the Empire few local officials were so rich or conscientious as to turn their backs on the opportunities that local office-holding presented. Later in the 1680s when no longer *corregidor* in Chayanta, Bóveda can be found still busily mining there, in a partnership with one don Lorenzo de Narrihondo y Oquendo, a prominent figure of the day in the Villa Imperial.[7] And in the last glimpse that the documents provide of him, Juan Antonio de Bóveda is still in Ocurí at the close of the decade, now in charge of the mines and the two well-equipped refineries that his brother-in-law operated there.[8] The intertwining of public and private affairs that Bóveda's career shows also marks the lives of other relatives of Antonio López. It was a combination that stood to do well by the *maestre de campo* and his family members—administrative power at the service of private interest.

Sometimes the links between the private and the public were less direct, but none the less potentially useful. Don Joseph de Bóveda y Saravia, for instance, another brother-in-law, provided financial guarantees for a newly appointed *corregidor* of Lipes in 1677, one *sargento mayor* don Joseph de Araujo y Gayoso, who was possibly also a Galician.[9] Two years later, don Joseph, now a *maestre de campo*, himself became López's agent in Lipes, taking over the running of mines and business there from don Juan de Velasco, the husband of one of López's daughters.[10] Or again, doña Agustina de Bóveda y Saravia, a sister-in-law to López, was the wife of Francisco de Arrazola, a magistrate of the Santa Hermandad, or provincial constabulary, for the Potosí and Porco area. The office was not a prominent one, but to have a relative in it could only have benefited López, especially in his mining projects at Porco.[11] Lastly, from his wife's family, López could call on a priest in the person of the *vicario* don Francisco de Bóveda y Saravia, trained in theology in Lima. Large families, such as the Bóveda y Saravias were in López's time, often counted a priest in their number. To have a direct connection to the church was useful, for both sacred and profane purposes. But it is striking to find don Francisco, despite his holy orders, working for López as a mining manager in the mid-seventies in Potosí. Here was another sort of authority, perhaps, that the *maestre de campo* found it profitable to apply to his only too earthly enterprises.[12] López, though, ultimately decided to make more conventional use of his spiritually-qualified brother-in-law. In 1677 he established a chaplaincy fund for him, worth 300 pesos annually. This sum was to pay for fifty masses to be said each year by don Francisco for López's soul, and those of his wife, daughters, father, and ancestors.[13]

Besides these immediate relatives in his wife's family, López could and did call on a large and sometimes confusing variety of other kinsmen occupying influential positions in and around Potosí—people bearing various combinations of the names Quiroga, Losada, Osorio, Rivera, and Sotomayor. With the exception of don Benito de Rivera y Quiroga, the leader of expeditions through the lowlands forests, who was López's nephew, it is not clear exactly what relationship these men had to the *maestre de campo*. But there can be little doubt that anyone with Quiroga among his surnames was connected by blood to Antonio López in some way. And of the closeness of the Quirogas and Losadas *fray* Felipe de la Gándara, the seventeenth-century historian of Galicia, exclaimed that it was so great by the 1500s "that they seemed one and the same family."[14]

The most apparent of these influential Quirogas was don Cristóbal Quiroga y Osorio, who first comes to light, already an honorary Captain, in the late 1660s.[15] In July of 1670 Viceroy Lemos named him *corregidor* of the province of los Lipes. Whether Antonio López had any hand in this appointment is an unanswerable question; but Quiroga's presence in los Lipes as regional administrator was certainly useful to López as he and don Alvaro Espinosa Patiño bargained with the mine owners in San Antonio over the terms of the driving of the great adit.[16] The normal span of office for a *corregidor* was three years, and indeed the last reference to this spell by Quiroga y Osorio in los Lipes comes in July 1672.[17] In 1678, he was still in (or back in) the province, this time as a private agent (and a *maestre de campo*) sent by López to attend to mining, trade, litigation, and the like.[18] He took over these duties, no doubt, from don Juan de Velasco, and they were to pass in the next year to don Joseph de Bóveda y Saravia. In 1680 Quiroga was still, nonetheless, representing López's interests in San Antonio del Nuevo Mundo in the lawsuit that had by then developed between López and don Alvaro Espinosa over payment of the costs of the adit in San Antonio.[19] By December of 1682 he was once more *corregidor* of the province of los Lipes, and specifically named, besides, as the superior magistrate of mines there.[20] By this time he was also producing silver in his own right at a site in that province called San Pablo, where he rented mines and a refinery seized for debt by the Treasury.[21] After then don Cristóbal Quiroga y Osorio disappears from view. By 1700 he was certainly no longer alive. But he had left a son in Potosí, yet another priest for the greater family, in the person of Dr. don Joseph de Quiroga y Osorio. In 1690 this young man, still in minor orders and studying for the full priesthood, became the beneficiary of the chaplaincy fund created in Potosí for him by one doña Isabel de Merlo, widow of the *maestre de campo* Domingo Vázquez del Villar, who had long ago, in the 1660s and early 1670s, served Antonio López as his administrator at Laicacota and Berenguela. The second and third patrons of this *capellanía* were none other than Antonio López himself, and his son-in-law, Captain Miguel de Gambarte.[22] Thus did the ties linking family, master, and servant intertwist and thread through time.

Another Quiroga closely associated with Antonio López and also with influential positions in local administration was don Diego de Quiroga y Losada, yet a further bearer of what came to seem the standard and indispensable military titles for men in López's immediate circle—Captain, and then after the mid-seventies, in don Diego's case, *maestre de campo*. He first makes an appearance on the Charcas scene in 1675, supervising work on the adit that Antonio López was then driving at Berenguela in the province of Pacajes.[23] Two years later, the viceroy made him *corregidor* of Porco. One of his guarantors (*fiadores*) for this appointment was still another Quiroga now active in and around Potosí, the *maestre de campo* don Joseph de Quiroga y Sotomayor.[24] From 1683 to 1686 don Diego acted as interim accountant (*contador*) of the Treasury office in Potosí, after all three incumbents of the senior Treasury posts had been removed, and indeed imprisoned, for suspected wrongdoing.[25] And finally, no longer a mere *maestre de campo* but a full *general*, he became *corregidor* of Cochabamba in the late eighties (possibly in 1687), from which position he reported on the history of the search for the Gran Paitití conducted by Antonio López and don Benito de Rivera y Quiroga.[26]

The don Joseph de Quiroga y Sotomayor who in 1677 provided a guarantee of the performance and probity of don Diego de Quiroga as *corregidor* of Porco was himself the holder of offices of some consequence. The first of these was the post of *protector general de los naturales* in Potosí, which he filled between 1669 and 1673, and possibly longer.[27] The *protector* was an official appointed by the colonial administration and paid a salary from the Treasury, with the task of defending Indians against abuses by officials, both Spanish and native, and by employers.[28] On don Joseph's performance of this charge there is no conclusive evidence. But since the *protector* in Potosí had authority over Indians who had been brought in for mine labor by the *mita*, it is an obvious possibility that he was in a position to influence both the supply of this draft labor to silver producers, and the outcome of any complaints that Indians might make about their masters in mining. So, in this case once again, Antonio López stood to benefit from another Quiroga's exercise of a function that impinged directly on his productive activities.

Don Joseph was a mere *sargento mayor* in the early seventies; but by the end of the decade he, too, had been raised to *maestre de campo*. And in 1679 the ruling viceroy, Archbishop don Melchor de Liñán y Cisneros, made him *corregidor* of Pilaya and Pazpaya—the position occupied by Antonio López's brother-in-law, don Juan Antonio Bóveda y Saravia, at the beginning of the decade. So once more the region where López maintained his great country estate would seem to have been in safe family hands.

Other examples of office-holding by Antonio López's known and presumed relatives might well be found. But, with some res ervations, the instances that have been given here add up to a good case for supposing that López could exercise through his rel atives, at least from about 1670 onward, and even more effectively in the 1680s, considerable influence on official action in the places—Potosí, Chayanta, los Lipes, Pilaya and Pazpaya—where he had his main productive enterprises in land and mines. And, this being the case, even if he had wanted to hold some office himself, for practical purposes he did not need to do so.

The common pattern that the careers of his relations followed is one of alternation between public and private life. A man would typically become *corregidor* of an area for three years, then in a private capacity switch to some economic activity—usually mining, given the possibilities of Charcas—for a few years, working for himself and often for López at the same time; and then slip back into public life for a while as a government official. The movement appears almost cyclical. The spell of public office perhaps charged the individual with authority and status that could then be advantageously deployed for private economic ends; but that reserve of authority eventually would become depleted, and another term in office was then needed to replenish it. When honorary military rank was added to civic office holding, as if often was, and with promotions, the standing of the man in question can only have been further boosted. The whole sequence seems to add up to procedures that stood to benefit the individual, and

Antonio López de Quiroga, greatly; government, naturally, may well have suffered.

The reservation that has to be placed on this argument, on the other hand, is the difficulty of seeing the system that has been suggested in actual operation. There is no positive evidence to show, for instance, that López was able to buy land especially cheaply in Pilaya and Pazpaya because one Quiroga or another was *corregidor* there; or that the terms that he and don Alvaro Espinosa extracted from the mine owners of San Antonio del Nuevo Mundo were more advantageous to the partnership than they would have been if don Cristóbal de Quiroga y Osorio had not been the supreme local official at the time negotiations were taking place. On the other hand, it is scarcely conceivable that López failed to benefit from his widely spread family contacts. And it can hardly have been a mere random outburst resulting from a personal grudge when the *general* don Gregorio Azañón y Velasco, a *corregidor* of los Lipes who was not a López man, complained in 1692 that the *maestre de campo* was "a powerful man who has held my predecessors almost at his command and disposition."[30]

Again though, little evidence exists on just *how* the Bóvedas and Quirogas surrounding Antonio López came by the offices they held. If indeed, as is suggested here, López set out to build around himself a network of kinsmen in public positions, it would be useful to show that he was able to engineer their appointments, perhaps, for example, by a well-timed donation to the Treasury or a discreet gift to a viceroy or high-court judge. Not surprisingly such evidence does not leap from the documents. And the best that can be proposed is that his evident acquaintance, and possible familiarity, with the Viceroy Count of Lemos was influential in the appointment of don Cristóbal de Quiroga as *corregidor* of los Lipes, or the naming of don Juan Antonio de Bóveda to the same post in Pilaya and Pazpaya; or that López's acting as business agent in Potosí for Viceroy Liñán y Cisneros, who had previously been Archbishop of Charcas, probably had something to do with don Joseph de Quiroga y Sotomayor's finding himself *corregidor* of Pilaya and Pazpaya in 1679.[31] Some circumstantial evidence is in place, then, to show that strings connected Antonio

López to the highest authorities in the realm, but he cannot be seen pulling them. On the other hand, there is not the slightest doubt that, however the situation came about, men from López's immediate and extended family were prominent in Charcas from the early 1670s onward not only for their private activities but for their public functions as local administrators. And Antonio López would have to have been a fool (which he manifestly was not) to fail to take advantage of the benefits that such a situation offered.

Mainstay of Potosí

In the 1670s López had poured his energies and investments into mining ventures outside Potosí, and many of these, particularly the mines and refineries in Chayanta and San Antonio, continued to prosper, well on into the 1680s. But this new decade was above all a time in his life when he turned his attention back to the Rich Hill of Potosí, reapplying to it the techniques that had first served him well in the Amoladeras system in the Hill, and then had been tested and perhaps improved elsewhere in the district. The essence of these techniques was, of course, driving deep adits to undercut previous workings, and by so doing drain and ventilate them.

López's interest in opening new adits in the Rich Hill, or in some cases of extending existing galleries, had grown by the mid-decade into a grandiose plan to drive tunnels straight through the structure at different levels. A survey done in 1686 by the official inspector-magistrates of the Hill showed that he had by then started work on two such adits: one beginning low down from inside the Amoladeras group, and the other from a long-worked section of the Hill, higher up on the east slope, called the "Polo site," or "*Paraje de Polo*" after a sixteenth-century miner who had worked there.[32] By the time of the survey, both of these reportedly had already intersected veins of good ore.

Three years later López had five adits in active progress in the Hill (in addition to seven others driven in the past, and still, apparently, usable or in use).[33] To establish a full legal description of them (and, no doubt, to arm himself with incontrovertible evidence of the great merit of these new undertakings), he petitioned the chief magistrate of mines in Potosí to carry out an official survey of all five. This was done on March 11 and 12, 1689 by the magistrate himself (the *alcalde mayor de minas*), Captain don Fernando de Noriega Rivero, accompanied by two inspector-magistrates, López's own chief mine supervisor, or *minero mayor*, Blas Míguez, and three other *mineros* in López's employment.

The inspection started with the "socavón de los chiles de la Amoladera"—the adit driven to cut across the Hill from the lowest levels of the existing Amoladeras workings. Using a compass, the inspectors found that the direction of the adit was east to west; and measuring with a cord from the iron grill that guarded the entrance, they found the tunnel to be some 580 yards long. At 90 yards from its face a transverse drift had been cut for over 200 yards to give access to a large vein of ore. No fewer than 65 veins in all, however, had been crossed by this adit in its progress to date, and 240 Indians were at work extracting ore from them. The inspectors thought, nonetheless, that 2,000 workers could easily be put to good use in the ore deposits revealed by this adit.[34] Its cross-section of almost seven feet of width by five feet six inches in height would certainly have been large enough to allow many men to come and go without difficulty.

The next day, March 12, 1689, found the inspecting team visiting López's other four adits in the Hill. One of them, the "socavón de Jofre," named after the miner Jofre Ibáñez who had made a small beginning on the gallery in the 1620s, struck into the Hill from a section called Berrío. This was on the east slope, slightly above the level of the Amoladeras. The intention here was to run the gallery westwards to intersect at great depth two of the earliest worked veins in Potosí, the Veta Rica and another called Centeno. The surveyors measured the adit at 660 yards. In that distance it had intersected five silver veins, though not yet, apparently, reaching its target. Twenty Indians were at work in the adit, cutting ore for López from the veins. A hundred, though, could well have been occupied in the same task, so the inspectors thought.[35]

Also at the Berrío site was the entrance to another of López's adits, called simply "Berrío," whose direction was toward the southwest. It was now 460 yards long, crossed three veins, and occupied twenty Indians.[36]

Above the Berrío area on the same east slope of the Hill was the Polo site, where the adit that López had had in progress in 1686 now ran 730 yards into the Hill, well on its way toward its aim of piercing all the way through to the west slopes. Six laborers were at work on the adit itself, though none in the three veins that it had so far crossed. The inspectors thought that 200 men could be put to work in those veins without any difficulty at all.[37]

Finally there was the "socavón del Rey," oriented northwest to southeast, and 460 yards in length. The starting point of this gallery is unclear. Three veins had been revealed by the tunnel to date; but only three Indians were at present working on it.[38]

So, all in all, the five adits that López had under construction in the Rich Hill in 1689 amounted to just over a mile and a half of tunnelling (2,890 yards)—a considerably greater length than in any of his projects out in the district. Two of the five, the Polo and Amoladeras galleries, were intended to traverse the Hill on an east–west axis, but at different heights. The other three were either directed toward known ore bodies (for example, the Centeno and Rica veins), or were started from sectors of the Hill where successful mining had taken place in the past.

These schemes for the Rich Hill drew admiring comments from witnesses to reports on the *maestre de campo*'s services that he caused to be compiled late in 1689 and in 1690. His plan to drain all the major mines of the Hill with his adits was "an apparently impossible undertaking," according to one of them. It stood some chance of success, nonetheless, because of López's willingness to lay out his wealth on projects of this sort.[39] Along the same line, the *corregidor* of the Villa Imperial, don Pedro Luis Enríquez, remarked

Some refiners have been seen in this realm to work the mines with success; but only in the *maestre de campo* have I seen anyone put back all his fortune into mining operations, adits, communication galleries, and drainage; and by such special service he has made himself unique.[40]

López would have been happy with this opinion, coinciding as it did with his own observation about himself that he had not "aimed to store up wealth, but to serve his Majesty in the increase of his royal *quintos.*"[41] Nor would he have quarreled with the judgment offered by don Juan Araujo, the man whom he had made chaplain of one of his refineries in Potosí, that "No monarch had had such a vassal." Though López could have retired in ease to his country estates, he had not done so, "but moved by the generous and liberal spirit that God gave him, he had desired nothing more than to work in mines"—and all, of course, to increase the king's royalties.[42] Happy he doubtless was, too, when one of his *mineros,* don Ignacio Pardo de Figueroa Montenegro, a fellow Galician, pronounced in 1690 that the lower levels of the Rich Hill were virgin, and unworked only because of the flooding which López's ambitious adits would now remedy. In these drainage efforts, and in mining itself, López, by Pardo's estimate, employed as many Indians in the Hill as all the other refiners and miners of Potosí put together. [43] One of the royal scribes of the Villa Imperial went still further: López, through his payments for wages and materials in Potosí, "sustains a third of the people living in this town."[44] A Franciscan ally declared that López had "been undeniably the most important and industrious *azoguero*" ever to work in Potosí.[45] Even a man who had once said of himself, as López did in 1679, that "it is public knowledge in these realms that only the valor and heart of the *maestre de campo* Antonio López de Quiroga keeps them on a firm footing"[46] could have found some satisfaction in such praise.

Whether López ever managed to finish his adits through the Rich Hill is not clear; the lack of reports on what would have been an unprecedented feat may well mean that he did not. Not much is recorded in general, in fact, about the mining and refining activities of his last few years, which suggests that finally his appetite for new ventures had waned. But some, at least, of what was already in place continued to operate. In 1690 he still had nine ore milling machines running on the *Ribera* of Potosí, and his refineries at Ocurí and San Antonio were also still active.[47] The Potosí mills were undoubtedly still processing ore from the

Amoladeras workings, which seem to have been the most productive among López's array of mines in the Hill, to judge from the inspectors' observation in 1690 that he had 240 laborers there, working the sixty or more veins that the adit had cut. Though a manuscript of 1695 hints that he had disposed of three refineries in Potosí by then,[48] in the following year he very probably still had at least one functioning *ingenio* in the Villa Imperial.[49] And to the very last days of his life he continued to rent from the Treasury the mill that had once belonged to Ignacio de Azurza in San Antonio del Nuevo Mundo.[50]

LAND AND TRADE

By some reports, the quality of the ores that López was refining had fallen severely by 1690.[51] This was one reason given for his heroic efforts to create new underground works: he hoped that these would bring to light fresh sources of profitable ore in the Rich Hill. Another means by which he tried to meet the high costs, and scant returns, of refining low-grade mineral was, by some witnesses' account, through diverting into mining his income from sales of agricultural goods in Potosí. Now, while it is uncertain how far he could have met his mining costs in this way, given their enormous size, this assertion does raise the interesting question of what use López made of his rural properties. The question is made all the more intriguing by the fact that, in contrast to his mining interests, which if anything tended to contract in the nineties, his land-centered activities did no such thing. On the contrary they give indications of expansion, and that from an already substantial base.

The hub of all López's landholding was the *hacienda* of San Pedro Mártir in the province of Pilaya y Pazpaya, which he had bought in 1658 and held to the end. In that same fertile and well-watered valley region south and east of Potosí he also acquired, sometime in the sixties, an *estancia* named Culpina.[52] And by late 1670 he had enlarged his possessions in this area with another property, described as *hacienda y estancia*, called Ingahuasi.[53] It was, apparently, these three pieces of land that formed the estate of several hundred square miles that López thought an appropriate territorial foundation for the title of Count that he so avidly sought in the seventies.

López's interest in land was to remain focussed on Pilaya y Pazpaya throughout his life. Only once does he seem to have bought a piece of agricultural land anywhere else. This was an *estancia* called Chaquilla, for a half share in which he paid 2,500 pesos in 1682. The property lay about 30 miles from Potosí, south of Porco on the road to los Lipes, and not far from the Indian towns of Tomahavi and Yura. Nothing besides the land itself is mentioned in the sale document or included in the price, except the right to the labor of half the permanent Indian workers (*yanaconas*) on the property. How many of these there were is not stated.[54] A little less than six years later, López sold this property again. Somewhere in the interim he had acquired the other half of Chaquilla, and the entire holding had clearly been much improved.

The term *estancia* usually implied livestock raising, and this is obviously what López had gone in for in a substantial way here. The single most valuable item in the sale contract of February 20, 1688, was 3,680 llamas, assessed at 4 pesos and 2 reales each, for a total value of 15,640 pesos. Besides these were 231 cattle, including 18 bulls and 51 breeding cows, altogether worth 645 pesos. A few tools, and small amounts of food obviously for consumption on the farm, were the other listed items. The land itself, with presumably a few buildings for workers and managers, and certainly a chapel, since its decorations were separately listed at 650 pesos, was appraised at 9,200 pesos.[55]

López's purpose in buying and developing this breeding land on the road between Potosí and los Lipes may well have been primarily to supply his mining operation at San Antonio del Nuevo Mundo with llamas for freighting. He would also have been very well placed, naturally, to provide men plying the carrying trade between Potosí and San Antonio with the animals they needed for their pack trains. From the cattle, one product might have been milk for cheese-making; but a more likely fate of the animals was early slaughter for jerky, hides, and tallow—all of them basic raw materials for mining operations in Charcas. Whatever the details, it is clear enough that at Chaquilla in the 1680s López created a livestock center of some considerable size. Why he sold it after only six years he does not say. Perhaps, despite his renewal of work in the great adit of San Antonio late in the eighties, he interest was mainly fixed by then on projects in Potosí, to which Chaquilla could not contribute greatly.

At the same time, attractive opportunities seem to have arisen in the hospitable province of Pilaya and Pazpaya that may have inclined López to shift his attention to his existing estates there. Only three months after selling Chaquilla he moved to increase his holdings in the southern valleys. The occasion arose when lands belonging to the family of don Alvaro Espinosa Patiño, López's recalcitrant partner in the San Antonio adit, came on the market. Espinosa had died early in 1687; and a year later his heirs and executors sold to Antonio López two pieces of wine-producing land in Pilaya y Pazpaya. One was named Carapari, and was evidently under cultivation; the other was abandoned and nameless. Both were near López's existing lands in the area. The sale included eleven or twelve black slaves, a house, a chapel, and other unspecified accouterments, for a total of 13,200 pesos.[56]

This purchase brought López's holdings in Pilaya y Pazpaya to five—San Pedro Mártir, Culpina, Ingahuasi, and now Carapari and its unnamed neighbor. But several more pieces of land were to be added in the decade that remained to him. By the end of his life he had further gathered in the province another *hacienda* (Los Duraznos), two vineyards (lands formerly owned by don Diego de la Presa and Rodrigo Sánchez de Caravajal), an undefined unit called La Cortiduría, and two pieces of "tierras," one called Sacaro, and the other simply "de Pilaya."[57] The total area of all these is impossible to calculate, for lack of measurements. But given that López had already claimed ownership in the 1670s of hundreds of square miles of Pilaya y Pazpaya province, the likelihood is that his holdings in their final form were substantial indeed. A minor pointer is that San Pedro alone in 1696 had eighty-four black slaves, both men and women; and by López's estimate, that piece of land alone was worth 150,000 pesos. It was "much improved," he said, in comparison with when he had bought it, long ago in

1658, for 52,000 pesos.[58] True to his usual pattern, López seems to have put in family members to run at least some of these rural concerns. One Antonio López de Quiroga y Valcárcel was the steward, or *mayordomo*, of Ingahuasi until his death late in 1692, when he was replaced, at any rate temporarily, by don Pedro Diez de Quiroga.[59] This same man, along with one don Benito de Boada y Quiroga, was empowered to run San Pedro Mártir in 1695.[60]

However much López may have emphasized to the crown in the 1670s that possessing large estates was an essential condition for holding a title of nobility, and in so doing, might seem to have been implying that for him the true worth of land lay in its being a traditional attribute of lordship, nonetheless not the slightest doubt can exist that he also saw his properties in the coldest of commercial lights. Hardly any hint is to be found of his farming production in the fifties and sixties, except that he bred mares at Culpina.[61] But by the early seventies, the lands in Pilaya y Pazpaya were sending tens of thousands of pesos' worth of goods to Potosí every year. Wine was the leading item. From 1672 to 1680 an average of over 1,700 *botijas*, or goatskins, of it arrived yearly. And between 1681 and 1689, the average went up to some 3,150 *botijas* annually. López's average income from wine sales in Potosí over the whole eighteen year period was 18,392 pesos annually. A very small amount of brandy was also brought in occasionally. After wine, cattle products were the main goods sent to Potosí from San Pedro and its companion holdings: jerky (1,280 *quintales* a year on average from 1672 to 1689, worth 5,596 pesos annually); tallow (651 *quintales* on average, worth 5,161 pesos); leather (2,511 pesos' worth yearly); white fat (228 *quintales* a year, worth 1,542 pesos); black, burned fat (115 *quintales* yearly, worth 685 pesos); and from time to time a little maize and *chuño* (dried potato). Taken together, sales of these things in the Villa Imperial brought López a yearly average of 34,309 pesos between 1672 and 1689—somewhat less in the beginning, somewhat more later, as a result of the growing amount of wine sent to market.[62]

This was a remarkably large sum, equal to about 1 per cent of average yearly silver production in Potosí and its district over the same period. But the indications are that it was substantially exceeded by López's shipments to Potosí in the late nineties. The gross retail value of the farm products that he placed in the Villa Imperial in 1696–97 was estimated to average 47,108 pesos a year. The entire increase resulted from growth in arrivals of wine; at 36,575 pesos in annual value, it was running at almost double the average of sales in the seventies and eighties. The other lines were much as before, or even below the previous averages: tallow at 5,818 pesos a year, jerky at 3,550 pesos, fat at 756 pesos, and leather at 360 pesos.[63]

Some of these items may not have been for sale, but, rather, for use in López's household and in his mines and refineries. This, at least, was what he argued in an attempt to reduce the amount of sales tax that the Treasury wanted him to pay for those two years. On the other hand, the collectors were sure that what he had brought to Potosí had been sold.[64] In the case, however, of the products sent from Pilaya y Pazpaya in the seventies and eighties, nobody, López included, denied they were commercial goods, brought to be sold in Potosí. López maintained a store yard, or *cancha*, from which they were distributed, under the supervision of a *canchero*, Pedro Marino de Lobera, who received a salary of 1,500 pesos a year from him.[65] Since López does not seem to have drawn on his *cancha* for supplies to his mines and refineries in Potosí, and obviously would not have bought on the open market necessities that his own lands could provide, it is possible, if not probable, that in the seventies and eighties his estates in Pilaya y Pazpaya sent to Potosí not only what arrived at the *cancha*, but also much else that went directly to his *ingenios* and underground workings: leather for containers, rungs of ladders, binding and reinforcement, hinges, clothing, and footwear; fat for lubricating wooden machinery; tallow for candles; jerky as food for employees.

This may have been the most useful service that his estates provided for his silver producing enterprises: a direct and reliable supply of necessary raw materials. Some people at the time thought López especially praiseworthy because he had been willing to use earnings from his land to support his mining ventures in times of difficulty.[66] But there must have been limits to what was possible in that respect, since sales of 30,000 or even 40,000 pesos' worth of farm goods a year in Potosí, however impressive in themselves, would not have gone far toward covering his outlays during the seventies on adits and other works, when the expenses in question ran to hundreds of thousands of pesos for each venture. Possibly, also, his agricultural income would have served as good security against loans, or as a reliable source of interest payments on borrowed money. But evidence is totally absent to suggest that he used it in that way, or even that he borrowed at all.

Antonio López, then, was remarkable as both a farmer and miner. He left no detail on how he ran his estate; but it is plain that in one important respect, at least, his approaches to both mines and land were the same: a penchant for enterprise on what was, by local standards, an exceptionally large scale. Possibly there were other landowners in seventeenth century Potosí whose properties spread wider than López's, but if so, they still remain to be discovered. Broadly speaking, also, rationality marks both his mining and his agricultural work. What evidence there is points toward a determination in López to bring abandoned or underused land into production, and to concentrate on two sorts of goods: items that were essential for the processes of silver production, and that therefore could either be sent directly to his own mills and mines, or find a ready sale among other miners; and other goods for which a reliable general demand existed in Potosí. Preeminent among these was wine. It is a little unexpected to find the gritty *maestre de campo* emerging as an ever larger producer of wine in the late seventeenth century. Perhaps his inclination in this direction, which is particularly visible in his final decade, was the result of a yen to turn to the gentler rhythms of the country. But just as likely a motive is a sharp assessment on López's part that, however unpromising the prevailing trend of silver production might be, wine was still something for which the Villa Imperial had an eager thirst. If Pilaya y Pazpaya could make wine, as the past showed it could, then let it rise to that demand.

Commercial awareness and an active interest in buying and selling, of course, went far back in López's life—first to the trading that may have preceded his arrival in Potosí, and then, if

Arzáns is to be believed, to his early importing and retailing in Potosí. This was followed in the fifties by his shift into the more specialized business of *avío* and silver brokerage. In 1679 he was still described as an *"azoguero* and *aviador* in this town and province, and a *mercader de plata* in the mint here,"[67] though there is little to indicate that he was doing much in the way of silver trading or giving *avío* by then, or that he resumed such activities later. On the other hand, in his later years he most certainly did conduct an energetic business in long-distance import of goods into Potosí.

The most striking example of this was his marketing of coca from Cuzco, far away to the northwest. From time immemorial, the leaves of the coca bush had provided the native people of the Andes with a pleasant and mild narcotic. Their habit was (and still is in many places) to chew the leaves along with a little ash, which helps release the active alkaloid from the plant tissue. In Inca times, only the nobility had been allowed to use coca; but such rules disappeared with the conquest; and the mass of the population took to the habit. Pedro de Cieza de León remarked on it in the 1540s. "Everywhere in Peru people . . . keep this coca in their mouths, from the morning until they go to sleep, without removing it. If Indians are asked why they always hold a portion of leaves in their mouths . . . they say that it makes them feel little hunger, and [at the same time] strong and vigorous."[68] Arzáns quotes this passage in *Historia* early in the eighteenth century, and says that nothing has changed in the Indians' opinion of the virtues of coca. "And so it is," he adds, "that no Indian will go into the mines, nor to any other sort of labor, such as building a house or working in the fields, without putting a coca in his mouth." Furthermore, native mine workers thought that the quality of the ore they were working would fall if they did not use coca.[69]

The demand for coca, then, in Potosí and elsewhere, was great; and Spaniards, rapidly recognizing a good business prospect, were quick to set up commercial plantations in the valleys on the inland slopes of the Andes. Arzáns remarked that in his day the only place that coca was raised was in such valleys, or *yungas,* lying almost 300 leagues from Potosí.[70] This seems surprising, since modern coca raising takes place in *yungas* much closer to the Villa Imperial. But it may explain why Antonio López had, apparently, to go as far as Cuzco, which is some 600 miles, or 200 leagues, northwest of Potosí to buy coca. It was, probably, from the Amazonian slopes of the mountains to the north of Cuzco that this coca came.

He first turned to this trade, so it would seem, in 1678, with the purchase of the astonishing quantity of 12,000 baskets, or some 300,000 pounds, of leaves. At 6 pesos the basket, the order was worth 72,000 pesos. This coca had been grown on the *hacienda de coca* of Captain Martín Gonazález Valero, a citizen and alderman of Cuzco, and the purchase was handled in that city by a man who was to remain López's agent there for several years, Dr. don Joseph Vázquez de Castro, the *maestrescuela*, or canon charged with teaching divinity, of the cathedral of Cuzco.[71] The participation of such a ranking cleric in big business might seem surprising at first sight, and not only for reasons of worldliness. One Peruvian author of the early seventeenth century had roundly declared "All those who eat coca are sorcerers who speak with devils, whether they are drunk or not, and go mad. God preserve us. Those who eat coca cannot receive the holy sacrament."[72] And Arzáns, a century later, said much the same: "The Devil has seized onto this plant coca with such ferocity that it is very certain that when people chew it, it deprives them of judgment as if they were overladen with wine, and makes them see terrible visions, and devils appear to them in dreadful forms."[73] The learned canon must surely have been aware of the fearful spiritual hazards lurking in the baskets of leaves that he despatched to Antonio López in Potosí; but perhaps he calmed himself with the thought that his own involvement in these particular dealings would be beneficial, and that his sanctity would preserve those who finally consumed the leaves from infernal dangers. A more likely comfort, alas, was simply the profit he stood to make as middleman. Clergymen in the Spanish colonies were no less inclined to business than anyone else.

Over the next eight years, up to the time of Canon Vázquez de

Castro's death (probably in 1686), López sent him further sums totalling at least 68,000 pesos. Whether this entire amount was for coca is impossible to say, though some of it certainly was, since López agreed in June 1686 to pay 2,300 pesos in sales taxes on 17,000 baskets of leaves that Vázquez de Castro had sent him.[74] After then there is no clear indication that López continued in the coca trade; on the other hand, there is no obvious reason why he should have given it up.

Throughout the late decades of the century, also, López constantly despatched large sums of money to Lima. Sometimes these were amounts that he had collected on someone else's behalf in Potosí, acting as a business agent. His clients ranged from the Archbishop-Viceroy don Melchor de Liñán y Cisneros, for whom he gathered ecclesiastical income in Charcas in 1679;[75] to the monastery of Our Lady of Victory in Baeza, in Spain, one of whose members had died near Potosí, leaving possessions that López disposed of for cash, which he then remitted to the Peninsula;[76] to a Lima merchant on whose behalf López had sold Andean-made clothing in San Antonio del Nuevo Mundo.[77] Other remittances to the capital were sums that the *maestre de campo* sent apparently on his own account, though for what purpose can only be guessed, since the orders he gave his agents in Lima have not survived. Perhaps he was buying things purely for his own or his family's use. The 17,000 pesos he remitted to Lima in March 1673, for example, though a large enough sum, would not have gone far toward buying the dowry goods that his daughter Lorenza took to her new household in 1676.[78] Or it is quite possible that López still occasionally engaged in a little importing of overseas goods through Lima to Potosí. If so, however, it was no more than a sideline.

In long-distance trade in a quite different direction, though, he most certainly did participate in the seventies and later. Once he had put together a substantial area of land in Pilaya y Pazpaya, which he managed to do by 1670 or so, he looked around for a supply of cattle with which to fill his wide pastures. The source to which he turned was the province of Buenos Aires, a thousand miles to the south. In September 1670 he contracted to buy 1,200 head of cattle, three years old or more, to be delivered to the Ingahuasi estate by the end of April 1672.[79] It is, to be sure, not at all clear whether the animals ever actually arrived.[80] But in the nineties he certainly did receive almost 5,000 head of cattle ordered from Buenos Aires for Ingahuasi,[81] and others are likely to have arrived over the intervening years, since cattle were not to be had locally on such a scale.

López's business interest in Buenos Aires went beyond cattle. In partnership with his second son-in-law, Miguel de Gambarte, who, of course, had been convicted of engaging in some distinctly shady import dealings in Buenos Aires in the mid-1670s, López began in the nineties, if not before, to buy merchandise in Buenos Aires for delivery to Potosí.[82] Whether, as in Gambarte's earlier dealings, the goods in question were partly of contraband origin is hard to say for sure. Given the wide availability of European merchandise, in and around Buenos Aires, some of it transshipped through Brazil, it is at least likely that some of what López and Gambarte bought came from some other source besides the restricted Spanish shipping allowed into the River Plate. Likewise, there must be a distinct possibility that López and Gambarte used the Buenos Aires route to smuggle out of Spanish America silver that had not paid the due royalty levies. San Antonio del Nuevo Mundo, in its isolated southern position, was a notorious source of such contraband metal. It was, as the president of the La Plata court noted in 1678 when writing to Spain about Gambarte's transgressions, significantly closer to the Río de la Plata than other mining settlements in Charcas (the saving was 150 miles on the straight-line distance between Potosí and Buenos Aires of some 1,100 miles).[83] But just as attractive to merchants as the economies that the shorter journey offered was the sparsity of governmental presence along the route to the south. Tax evasion was easy, and that enabled traders to pay a higher price for unminted silver destined for Buenos Aires than for metal sent to Potosí. The difference in price was in truth so great that even the most conscientious miner at San Antonio must have been sorely tempted: 6.5 pesos for a mark of raw silver bound for the Villa Imperial against 8 or even 8.5 pesos per mark exiting through the Río de la Plata.[84]

Though the authorities could inform themselves fully of these machinations by merchants and silver dealers, it was another matter to do anything about them. A particular obstacle to enforcing the law, so the president of La Plata complained (and here López's political constructions are brought to mind) arose when men connected by some regional affiliation from Spain banded together in these maneuvers. "Especially when we come up against some business of countrymen and homeland (*"paisanos y patria"*) do we find that this [loyalty] prevails over all other respects of conscience and justice. . . ."[85] The president names no names, except Gambarte's. So whether he was thinking of Antonio López's collection of Galicians, already well formed by 1678, can only be surmised. The *corregidor* of San Antonio in the early nineties, don Gregorio Azañón, though admittedly no friend of López's, certainly implied strongly that the *maestre de campo* was then exporting untaxed silver southwards, or at least backing other miners who did so.[86] It is indeed difficult to imagine such an enterprising spirit as López, especially with Gambarte's sharp influence acting on him, managing to resist such opportunities completely.

If, though, Antonio López de Quiroga did permit himself a slip of this sort late in his career, it was an uncharacteristic lapse from what seems to have been a generally high standard of business conduct. López had qualities that gathered him enemies: assertiveness, implacability in pursuing what he saw as his rights under law, impatience with those less energetic, able, or daring than himself. But, as his repeated appearance throughout life as the agent for others' business shows, he was a man trusted by many, in Potosí and far from it, to handle their personal and commercial affairs. Nowhere is this more apparent than in his role of syndic for the Franciscan house in Potosí, and later for the whole Franciscan establishment in the province of Charcas. One member of the monastery in Potosí, *fray* Antonio del Puerto, wrote of López in 1689 that, over his thirty-seven years in the office of syndic, he had not been "like the other syndics, who have been content with the limited matters of collecting income and spending it;

but [rather], like none other, distinguishing himself by his devotion, fervor, and charity, [López] has donated excessive alms and temporal improvements to this monastery. . . ."[87] He had given much money, annulling debts that the monastery owed him, constantly sending gifts of food, presenting sumptuous ornaments for the celebration of the Mass, succoring passing friars with money and goods, and sustaining prelates in their visitations and peregrinations. Recently, having seen that the refectory, its antechamber, and other adjoining offices of the monastery were in bad repair, López had rebuilt them all at his own cost, adding wall panelling and moldings of cedar to the refectory and anteroom, donating tables and high-backed benches of the same wood, and even providing a new kitchen, with accompanying pantries and cells, for the greater ease and service of the community.[88]

While attending to the material comfort of the friars, though, López did not forget his own spiritual needs, which now, in his advancing age, must have been ever in the forefront of his mind. In anticipation of his own end he had a sumptuous vault built in the monastic church. Here he was to be interred, along with his heirs after him. The tomb lay before the altar of the Virgin of the Immaculate Conception, to whom, like many Spaniards of his century, he was greatly devoted.[89] He was, indeed, the steward of a brotherhood in Potosí dedicated to the *Inmaculada*.[90] For her chapel in the Franciscan church he had donated an altarpiece of gilded cedar,[91] an iron screen painted in green and gold, many cakes of wax, and precious decorations to adorn the image of the Madonna: crowns, veils, and rich vestments, some embroidered in gold, and others of Milan cloth. Still to be installed on the altar in 1689 was a frontal of solid silver, weighing fifty pounds, that López had commissioned. It was a sheet of silver on which decorated panels were raised in relief. In workmanship alone this object had cost more than 1,200 pesos. "No finer thing can ever have been seen in this town," said *fray* Antonio del Puerto.[92]

It was in this richly adorned setting of his own creation, an expression certainly of his spiritual devotion, but also a monument to his wordly success, that Antonio López came finally to rest in 1699. "As each man's life may have been led, so will his death be," wrote Bartolomé Arzáns de Orsua y Vela less than a decade later,

for death is a shadow that follows life as a shadow follows the body; and so to judge if a man died well, one should look to see if he lived well (for the soul of such a man will be blessed in life): there is no bad death if a good life preceded it, and the death of the sinner is always bad. Good in all things was the life of the *maestre de campo* Antonio López de Quiroga, and so, accordingly, was his death. He passed, then, into the repose of eternal life (so we may piously believe) in the month of April[93] of the year of 1699, being buried in San Francisco, his body lying beside that of doña Felipa, his wife.

It had been a good life too, no doubt, in López's own estimation, though perhaps by a measure different from Arzáns's conventional moralizing. The career and the written words of Antonio López, few though these regrettably are, suggest that for him a great good, possibly the greatest of all, was focussed action. To wrestle with and overcome Nature, using all the means that effort, concentration, organization, capital, and technique could supply—that became for him a supreme end in itself. From it might well then follow other rewards: personal wealth and power, esteem as a pillar of Potosí and the wider colony, a sense of service to the monarchy as a whole. But if these failed to come, as sometimes they did, then there was always new action to fall back on, a profound satisfaction and pride in throwing self and resources into renewed combat with rock, water, and ore. Those who got in his way, or simply seemed fainthearted, were roughly pushed aside. There is no striking goodness, in the sense of charity, in Antonio López. But for him, to do was good. And as a doer, he was not merely good, but the best.

CONCLUSION
LESSONS QUICKLY LEARNED
AND WELL APPLIED

SILVER PRODUCED

ALTHOUGH IT WAS ONE OF ANTONIO LÓPEZ'S FAVORED FRANCIS-cans who made the ringing statement that "Indisputably the *maestre de campo* has been the *azoguero* of greatest account and industry ever to be seen in this town,"[1] this, nonetheless, was a widespread opinion in Charcas by the late years of López's life, and one, moreover, for which much hard evidence lay at hand. Anyone who doubted it had only to go and count the *ingenios* along the *Ribera* of Potosí, visit López's great adits in the Rich Hill, or, if he had the patience and the necessary official authority, consult the records of *quinto* payments and mercury sales kept by the royal Treasury in the Villa Imperial.

These accounts show beyond doubt that throughout his career as a silver miner and refiner, López was responsible for the production of a steady and substantial proportion of the combined output of Potosí and its wider district. In the 1660s, the mercury sales record strongly suggests that his mines and mills yielded almost 13.5 percent of the total produced, with a marked increase from the first half of the decade (9.7 percent) to the second (18.6 percent), as his workings in the Amoladeras section of the Rich Hill grew and prospered.[2] In 1674, López received no less than 22 percent of all the mercury distributed by the Treasury in Potosí.[3] And while it may be unwise to put much weight on one year's figures alone, it is at least possible that this high proportion is indeed a reflection of his actual silver production at the time, as workings outside Potosí, especially in Chayanta in these years, were added to his ventures at the Amoladeras.

Early in the following decade, López's share of mercury sold was down from the high level of 1674, though very much the same as the average for the sixties. In four distributions totalling 4,986 *quintales* made in 1682 and 1683, he took 607 *quintales*, or 12.2 percent.[4] This may not seem a remarkable proportion of the total—except when it is considered that in 1682, for example, sixty-four others besides him bought mercury from the Treasury in Potosí, receiving an average of about 39 *quintales* each, and that no other individual bought more than 110 *quintales*.[5] If all this mercury was being used to process ore into silver, it can therefore be said that, other things being equal, López in the early eighties was responsible for about an eighth of the output from Potosí and the district, while nobody else could have laid claim to more than a thirtieth of the total.

Another, and perhaps more reliable, assessment of López's contribution to silver output over many years can be drawn from records of his payment of silver taxes at the Potosí treasury office. In 1690 the *factor*, or business manager, of the office calculated that López, between January 1673 and the end of June 1690 had brought in 6,208 bars of silver, produced in his own refineries from ores taken from his own mines, to be taxed with the royal fifth. The combined weight of all these bars was 918,613 marks, or 459,307 pounds.[6] Now the total quantity of silver taxed by the treasury in the Villa Imperial for that same period was on the order of 6,288,170 marks.[7] López's 918,613 marks come to 14.6 percent of this amount.

From the beginning of his concerted mining and refining activities in the early 1660s, then, until 1690, Antonio López seems likely to have been the producer of between a seventh and an eighth of the silver yielded by the mines of the Rich Hill of Potosí and the various other mining camps scattered around the great district at whose center the Hill stood. There is no record that anyone else among the sixty to seventy *azogueros* active at any given time during his career came even close to equalling him in either share or consistency of production. Nor, for that matter, is there any figure in the earlier history of Potosí who can be identified as Antonio López's rival in proportional or absolute output of silver. It was as Arzáns remarked: "he built some *ingenios* and bought many others until he had in the Villa alone eight ore-milling machines, a thing without equal since the celebrated *Ribera* was first constructed."[8]

Though nobody, as far as can be told, including López himself, kept a complete and continuous record of how much silver he produced over his entire career, a reasonable estimate of the figure can be reached by working back from the total registered output of the entire district in his time. From the beginning of 1661 to the end of 1698 (a month before his death), 13,061,591 marks of silver were brought for taxation to the Treasury office in Potosí.[9] If, as suggested above, Antonio López's mines and mills yielded a seventh to an eighth of this total, then his cumulative production over his career (ignoring any small quantities he might have refined before 1661, and whatever amounts he failed to register for taxation), lay somewhere between 1,633,000 and 1,866,000 marks of silver.[10] At 8 ounces to the mark, and the present (December 1987) silver price of about U.S. $6.70 to the ounce, Antonio López therefore produced between 88 million and 100 million dollars' worth of silver during his forty or so years as an *azoguero* of Potosí. Though enormous changes have taken place between the late seventeenth century and the late twentieth in the relative prices of goods, and indeed, of course, in goods available to be bought, nonetheless this conversion into modern terms gives some broad impression of how large López's scale of operations was. On such a scale, even in today's vast and industrialized economies, only a small fraction of producers could rank themselves high.

PROCEDURES

The question that immediately follows from a demonstration of Antonio López's preeminence among silver producers of Potosí of his own, and earlier, times is obviously: How did he do it? Though the most persuasive answers to this question are several-sided, one initial point—a point of departure—can be stated without hesitation. It is that, unless López's success is to be put down to luck, his very predominance in silver production means that he must have deployed exceptional personal abilities in organizing and achieving that production. Now in mining the effects of luck can never entirely be disregarded, though the greater the skill of the miner, the less he is at the mercy of fortune. But in general it is accurate to say that the opportunities presented by Potosí to López—in ore deposits, mining and refining techniques, access to labor, business structures, and political organization— were available to many others around him. If he took them and combined them into productive enterprises that yielded far more richly than other men could manage with their own schemes, then inevitably it must have been unusual qualities in himself that produced that result.

"The *maestre de campo* Antonio López de Quiroga," wrote Arzáns, "was, then, of Galician birth, of illustrious lineage, of singular graces, of settled disposition, of great truthfulness, and of very lively understanding. In desisting, he was wise; in venturing himself, daring; in undertaking, spirited; . . . in advising, prudent."[11] It was the possession of such qualities, and their application to the economic, political, and social opportunities around him in different combinations at different moments of his career, that seems to have given López his edge in Potosí.

In his first decade of residence in Potosí, the prudent side of Antonio López is the most apparent. This was for him a time of settling in, of observing, of learning the ways and the business of the Villa Imperial. His Galician origin and the social standing he brought from Spain certainly aided him in these tasks, for it is likely that they provided him with an entrée to a group of fellow-countrymen established in and around Potosí, among whom there were a few figures of some, though not remarkable, stature; and it is equally probable that his origins in Spain in particular favored his acceptance into the Bóveda family as a son-in-law, which in turn gave him access to wealth additional to whatever sums he had brought with him to Potosí. Grace and an air of honesty (and there is little cause to suppose that the air was not also the reality), too, helped him set himself up in the town. Something, for instance, besides the fact that his father-in-law, Lorenzo de Bóveda, had been syndic of the Franciscans in Potosí must have inclined the friars to choose this newcomer to the Villa as Bóveda's replacement in 1652.

In business, those early years were for López a time of cautious watching, experimenting, and learning, as he moved during the first half of the 1650s from general trading, to *avío* of miners (supplying credit and goods on credit), and finally to full-scale silver brokering. He may have received special encouragement to enter on this course from don Francisco de Nestares Marín, and, having set out on it, he was certainly helped by Nestares's reductions of taxes on silver. Indeed, Nestares may have cleared López's way in a more general sense by removing many of the previous silver brokers from the mint in Potosí as part of his stringent reform of that institution. But that advantage, and the tax cuts also, were available for the benefit of anyone who wished to try his hand as a *mercader de plata* in Potosí after 1650. More than the recipient of lucky favors, López should be seen in these circumstances as a man alert to an opportunity that most others seem to have missed. In his quickness, however, he was not rash, but rather advanced circumspectly into the credit and silver business as the partner of someone, Juan de Orbea, already well familiar with it. It was only after some five years had passed, and shortly before Orbea's death in mid-1655, that López began to practice the full range of activities of a *mercader de plata*.

López's actions over the following few years are the most puzzling, and the most original, of any that he took during his career in Potosí. For in the late 1650s and early 1660s he diversified from *avío* of miners and handling the silver that miners produced, into mining and silver making himself. It was a strange course to take, for, as president of La Plata had written twenty or so years before, no prosperous man in Charcas had ever gone looking for mines. Rather than face the physical discomforts and the risks of mine work, the wealthy preferred to take advantage of mineral discoveries through the profits to be made from supplying *avío* to the miners.[12] López, by contrast, was evidently willing, and appar-

ently keen, if not to abandon the more comfortable and traditionally safer business of credit for the uncertainties of mining, at
least to add those uncertainties to his existing business of silver
brokerage and credit.

López's own explanation of this diversification—that his losses in *avío* had become so great that he had no choice but to try to
make money through mining—leaves much to be desired.[13] For
one thing, it is by no means clear that he did suffer any net losses
on his credit and brokering dealings in the 1650s. For another, it
suited his purposes of self-aggrandizement to emphasize his early difficulties and his heroism in vanquishing them. For yet another, he had had no firsthand experience of mining and refining,
and could hardly have had the confidence, especially if his existing business had proved so disastrous as he maintained, to expect
that he could succeed where local experts had failed. And lastly,
only a foolhardy and even desperate man would have looked for
financial salvation in silver production, given the long decline
that mining had suffered in Potosí during the half-century before
López took it up. And there is no indication that López was giv-
en to acting in foolhardy and desperate ways. Indeed, his very entry
into mining had quite the opposite character. Like his other
actions in the fifties, it was a slow and even tentative affair, beginning with two leases of royal mines in 1657, followed by the purchase two years or so later of a cheap and run-down *ingenio*.

To go by the available evidence, therefore, a more plausible
explanation of López's movement into mining than his own rather dramatic account is simply that he decided to try his hand at
silver making, the fundamental business of Potosí, in an inex-
pensive and typically cautious fashion. A further incentive for
him to do such a thing may have been that, through his *avío* transactions, he already had much money in effect committed to mining, some of which he certainly was unable to recover. Any means
of making use of this investment would be welcome. So, for example, when one of his clients, Bartolomé de Uceda, died, owing
López many thousands of pesos for *avío* of his *ingenio*, López, as
a principal creditor, was able to take over and use the mill by
early 1659 for his own purposes. Prudent experimentation at low

cost, coupled with a desire to extract some return from otherwise bad debts, may then have been what led López de Quiroga
into silver production a decade after he arrived in Potosí.

The experiment, however cautiously embarked on, was clearly an enormous success. For while López before the beginning of
1659 was unknown as a silver maker, by the end of 1661 he may
well have been refining 5 percent of the silver produced in and
around Potosí (to judge by the amount of mercury he bought in
that year).[14] The precise steps that he took to make this remarkable progress cannot be traced, for lack of evidence. But whatever they were, he learned from them quickly and well; for after
1661 he shows all signs of being unstoppable as an *azoguero*.

The mines that carried him within a few more years from this
promising start to a position of overwhelming dominance in silver production were the Amoladeras deposits in the Rich Hill of
Potosí. After Antonio López's first claim to a working in them
was registered in September 1661, he moved with great speed and
energy to assemble a group of adjacent mines in this sector of the
Hill. An element of good fortune can be seen here, in a gift of an
adit and mines received from doña Paula de Figueroa, daughter
and heiress of a well-established miner of Potosí. But most of
López's workings in the Amoladeras came to him through obviously planned registrations, in his own and relatives' names, of
previously abandoned workings. To these were added royal mines
in the Amoladeras that López was able to rent for nominal annual fees.

Just why López resolved initially to concentrate his efforts so
intensely on the Amoladeras deposits is impossible to say for sure.
This part of the Rich Hill had not been extensively worked before,
although at least once, earlier in the century, its high potential
had been noted, and López may well have followed up on reports
he had heard among the mining men with whom he had had financial dealings. There seems little doubt, on the other hand, about
the rich rewards that he received from his efforts in this set of
mines. Those rewards, nonetheless, were as much the outcome
of the technique of mining that he employed as of the natural
wealth of the deposits. The essence of that technique was to maxi-

mize ease of access to, and extraction of, the ores by driving a series of interlinked adits among them. While adits were by no means an unknown and untried device of mining in the Rich Hill of Potosí, no one had previously used them on the scale that López now adopted. Broadly speaking, miners before him had inclined, more or less, to the traditional and, so to speak, "natural" mining method of driving their tunnels wherever the vein of ore led them. Access galleries were of secondary concern. If a man could move along them carrying or dragging a sack of ore, then the gallery was acceptable, no matter how narrow, twisting, or steep it might be, or how long a carrier might take to reach the surface with his load.

At the risk of oversimplifying, it can be said that López's mining technique, as first exemplified in the Amoladeras, reversed the emphasis. First, good access to the ore bodies must be created, by means of wide, high tunnels, spacious enough to let men pass in opposite directions, even if loaded; once that was done, mining itself could go forward. And adits, of course, if well planned, could do much more than provide just access. If they were driven into the bottom of the region to be mined, and if they were given the proper, gentle slope toward the exterior, then entire groups of large workings could be drained through them. This may have been one of López's main aims and accomplishments in the Amoladeras, since the lower levels of the Rich Hill, in which these ores lay, were more subject to flooding than the higher reaches, where most previous mining had been conducted. Adits, finally, and especially systems of interconnected adits, greatly improved ventilation, with several benefits. Workers would certainly perform more efficiently and perhaps more willingly if provided with a decent supply of fresh air; heat (which is considerable in the lower reaches of the Rich Hill of Potosí) would be reduced; and candles and torches could be used in greater numbers, and would burn better, with an increased supply of oxygen.

Increased orderliness in the layout of underground workings, then, was the chief mark of López's exploitation of the Amoladeras ores in the Hill of Potosí. The source of this innovative rationalization is hidden from view. Perhaps it was some supervisor whom

López employed to manage his mines: later examples show that he took care to choose the most able of such men that Potosí had to offer. Or, it might be argued, improvements of this variety were exactly what might be expected from an intelligent and ambitious newcomer to the business, such as López was. Untrammeled by received practices and habits of mining, he may have been precisely the sort of person most likely to discern both the inefficiencies of past methods and the obvious solution to them. But whatever the origin of the scheme employed at the Amoladeras, one outcome of that venture is clear beyond dispute. And that is that López learned well and fast from its success. As, in the course of the 1660s, he cut and linked his adits to open up the rich ores of this section of the Hill, he was laying down the foundation of the rest of his mining career in the Potosí district. For what he practiced here he would apply elsewhere in the 1670s and 1680s, both in the Rich Hill itself, and in other mining centers, in the series of great adits for which he was to become famous in his own time.

The Amoladeras mines, furthermore, supplied not only the ambitious technical model for Antonio López's subsequent enterprises, but also, to judge by the available evidence, the capital that made applying that model possible. It was these mines, after all, that gave the ore that made him the source of almost a fifth of Potosí's silver (according to the mercury record) in the second half of the sixties. Nor was this all that he gained from the Amoladeras venture. His triumph here engendered a shift in attitude, in his approach to business. The caution of the first decade vanishes after the sixties.

What dominates now is the spirited and daring enterprise that Arzáns remembered. While never foolhardy, and scarcely rash (except perhaps in his spending on the pursuit of the Gran Paitití), López, after 1670, can no longer in any way still be seen as the methodical learner that he had been in the fifties. In his own mind, he was past all that. Now he was confident of holding the true formula for successful mining in the Potosí district. And events in the sixties provided strong reinforcement for this confidence, as his investments in large and ambitious underground works in Chayanta and San Antonio del Nuevo Mundo began to bear fruit. Porco, it is true, did not do as well; and Laicacota was frankly disappointing. But these were more than offset by successes, and did nothing to dissuade López that his strategy of unstinting investment in deep galleries (which indeed might be called "radical" in the literal sense that he aimed them at the base of large ore bodies) was the right one for unlocking the unused potential of the Potosí district. It was in the last year of the seventies, his great decade of expansion, that López so severely reproached his partner at San Antonio, don Alvaro Espinosa Patiño, for failing to throw all that he possessed into the realization of their ambitious designs for those mines. By hanging back, Espinosa had lost his chance to become the richest man in Peru. The criticism seems perhaps touched by rhetoric. But it is just as likely to have been literally meant. By this stage in his career, López may well have been unable to understand, or to tolerate, the caution of ordinary men, so ingrained was his own self-confidence. It was, indeed, in that same year of 1679 that he informed his king, in altogether matter-of-fact words, that the general opinion had it in Peru that it was only his courage and efforts that kept the colony on its feet.[16]

For a man so sure of himself, great business decisions were the stuff of life. If repeated experience showed that long, deep adits were what was needed to unlock the remaining silver ores of Charcas, then such adits must be driven, even though the same experience showed that they would be several years in the making, and would swallow two or three hundred thousand pesos before any return could be expected. The president of the court in La Plata caught López's spirit well, and with minimal exaggeration, when he reported to the king in 1682 that, from his great income, the *maestre de campo* had reserved for himself

only the vanity of having distributed it, and the desire to have yet more to distribute, by overturning and disembowelling mountains, some of them of living rock. It is a matter of special providence for the maintenance of so risky a profession [as mining] that there should exist someone of such dedication; for if there were not these spirits who place more

value on joyful hope, with which they invigorate their plans, than on
the safe operations that might [merely] sustain them, mining would have
been forgotten long ago. . . .[17]

Confident, even exuberant, enterprise, then, was the spirit that
drove Antonio López's actions in the last three decades of his life,
the outcome of lessons well absorbed and applied in the fifties
and sixties. The fact that he learned his central procedures early,
though, does not mean that he then became inflexible, unimagi-
native, or closed to beneficial additions to his established approach
to mining. His rapid application of blasting to many of his projects is in itself enough to show that he remained open to advantageous change. Precisely how much gain powder offered cannot
be calculated. But for López in particular, with his conviction that
deep, and therefore unusually long, adits were the solution to
Potosí's mining problems, the increased speed of tunneling provided by blasting may well have brought great savings. It seems
safe to say that without blasting the scale of his undertakings
in the seventies would inevitably have been far more modest
than it was; though on the other hand his having created the
Amoladeras system without the use of powder suggests that this
technological development was not absolutely critical to his
success.

Innovative in a quite different way was the network of friends
and relatives of López's in local high places that first began to
appear in the seventies, and developed from then on. While there
is little evidence to demonstrate how López might have set up a
structure of this sort, it is beyond belief that mere chance led
several of his brothers-in-law and other bearers of the surname
Quiroga to appear with some regularity as *corregidores* of areas
where he had mining or landed interests. The conclusion must
be that he had a hand in these appointments, through friendship
and other contacts with viceroys and lesser, though influential,
officials of colonial government.

Even in the unlikely case, however, that López had no part in
creating this particular piece of organization, there can be no doubt
that, when his mining career is broadly observed, organization,

indeed integration, emerges as its salient quality: integration of underground workings, by the use of adits, into units of hitherto unequalled size; integration of mining and refining in different regions into a whole through the use of relatives as administrators; integration of silver production with land ownership, through the use of agriculture to produce raw materials needed in mills and mines, and possibly also through the application of income from sales of agricultural goods to mining purposes. To this list might be added the integration of the traditionally separate roles of silver maker and creditor, since Antonio López, besides pursuing the formal trades of *aviador* and *mercader de plata* early in his career, seems, even when he had apparently abandoned these pursuits to a large degree, to have been his own source of capital. In that way, he can only have saved himself the considerable borrowing costs that most of his competitors had to bear. As a final and imposing symbol of Antonio López's urge to integrate, there stands his last, sweeping scheme to pull together the ore deposits of the Rich Hill of Potosí by means of adits driven through it from one side to the other. That this project was not, apparently, completed by the time of his death does not take away from the grandness of its conception.

COSTS REDUCED

In the generally downward track of silver production in the Potosí district during the seventeenth century, there is a levelling-off, a platform, between about 1660 and 1690.[18] This is exactly the period of Antonio López's greatest efforts in mining and refining, and it is obviously tempting to attribute the pause in the downward movement to those efforts. This, however, cannot be shown to be so. To have been the source of about a seventh of the district's total silver was a remarkable and unprecedented feat, but López would have had to dominate Potosí far more completely to stabilize output as a whole. This is not to say that the example of his energy and methods may not have moved other miners to imitate him, and that may have contributed to arresting the long, previous decline. Future research may reveal some such effect.

At the very least, though, it can be said as a certainty that López's own mining career was a contradiction of the dropping trend so well set in Potosí mining by the time he came to the Villa Imperial. If the general causes of this contraction were, as argued early in this book, the growing costs of securing labor and of extracting ore in comparison with conditions in the sixteenth century, then, to have succeeded as he did, López must have managed to reduce these expenses in his own case to some substantial degree.

It is in mining rather than in labor that he seems likely to have worked his economies. With one exception, there is nothing to suggest that he was able to find laborers at any lower cost than other miners had to pay. The most obvious way in which he might have tried to do this would have been to deploy his connections to seize a larger share of the *mitayos*, or draft workers, sent every year to Potosí than his contribution to total production would have warranted. His best opportunity for doing this (and this constitutes the exception just mentioned) would seem to have been when don Joseph de Quiroga y Sotomayor held the post of *protector general de los naturales* in Potosí in the late sixties and early seventies (see chap. 4). But, apart from that, there is no hint of his enjoying any special access to labor. He did, certainly, get allocations of *mitayos*, like any other active silver producer in Potosí, but he also hired large numbers of wage-laborers, or *mingas*, who were generally three times as expensive as the draftees, or even more. When the inspectors of mines surveyed the five adits that López was making in the Hill in 1689, they found 283 Indians working in them, of whom almost two-thirds (183) were *mingas*.[19] In his refineries, it is likely that very few draft workers were to be found, since it was the general pattern in Potosí to hire wage laborers for milling and amalgamating: they were generally more skilled and experienced than *mitayos*, and refining demanded knowledge. And outside Potosí, where López was particularly active, he was almost entirely reliant on wage labor, since, with the exception of Porco, mining centers out in the district never received allocations of draft workers. All in all, it would seem that far more than half of López's native labor force consisted of waged workers, while on average in seventeenth-century Potosí (or to be precise the first half of it) the division between *mitayos* and *mingas* was roughly equal.[20] So, if anything, López is likely to have faced relatively higher labor costs than the general run of miners and refiners around him. Offsetting this, on the other hand, may, if Arzáns is to be heeded on this topic, have been a higher productivity of labor. For just as López was known for his liberality to his stewards and mine managers, so he may have paid his laborers well. "What he most urged on his *mineros* and majordomos was that the Indians should be fully rewarded for their work; because, he said, if this were not done, God would take from him what He had given."[21] But no other evidence remains on this point, and its weight is impossible to assess.[22]

It seems, then, that it was mainly through lowering the cost of extracting ore that López was able to defy the trend of silver production in Potosí that had predominated since 1600. This reduction, of course, was precisely the sum effect of his strategy of attacking known ore bodies with new systems of low adits that bypassed the existing, and usually inefficient, steep shafts through which they had previously been worked. Integration is at work again here, in the tying together with a single stroke—the well-placed adit—of disparate mines. Simplification went together with integration. Difficult and contorted shafts were replaced by a single, direct gallery, permitting quicker, easier, and hence cheaper, extraction of ores. Of course, creating the simplification was immensely expensive, and hence risky. But it is a measure of López's enterprising qualities that once the advantages of simplifying had been demonstrated to him, at the Amoladeras site in Potosí, he did not hesitate to pursue the same effect vigorously elsewhere. Whether he ever decided against driving an adit at some particular mining site because it seemed to him that the outlay could not be recovered, is not to be known. To see whether he decided to proceed or to hold back as the result of some rational calculation, or of some intuitive judgment, would be immensely interesting. But the record does not permit of any such insight.

INVESTMENT

Behind much of López's integrative effort lay, then, heavy investment. It is impossible to estimate with any convincing precision how much this investment might have amounted to, all in all, in cash: though it would seem, if López's own round figures for his spending on adits have any validity, that on these alone he laid out at least between one and two million pesos.[23] And the purchase and equipment of his estates must have run to at least 200,000 pesos, given that his initial acquisition alone, of San Pedro Mártir, cost 52,000 pesos. So, despite the conspicuous frugality of his spending during his entry into mining in the sixties, in later decades he was prepared to match his outlays to the scale of his ventures. This was, indeed, an aspect of his behavior that profoundly impressed his contemporaries. Dr. González Poveda in La Plata exclaimed over his willingness to plough back into mining the earth a great proportion of the silver he had drawn from it. Other witnesses were consistently impressed by the lavishness with which López's refineries were equipped. It was common knowledge, one of them noted, that he "spared no expense, however great" in his quest for silver.[24]

These remarks, and López's own insistence on the necessity of a miner's committing his entire resources to his task, are, alas, but subjective evidence on the question of the *maestre de campo*'s propensity to invest. And there is not a great deal more to go on. An oddity in another part of López's behavior, however, may give a sort of negative confirmation of his inclination to devote his resources to mining to an unusual extent. This is his attention to charity. For all that one witness late in López's life enthused over the aid that Indians and monasteries in the Villa Imperial could find "in the compassionate bowels of the said *maestre de campo*,"[25] and another declared that López was "much loved" in Potosí both for his liberality with wages and for they many alms he distributed,[26] there is in reality little to indicate that he was generous in proportion to his wealth. Grateful as the Franciscans of Potosí may have been for his help with business and for his material support, he really did no more than repair or embellish parts of their buildings. This was a far lesser thing than the creation of a whole foundation that a prosperous Mexican miner had managed early in the century, giving 100,000 pesos to the Jesuits to build and endow their house in Zacatecas.[27] It fell far short, too, of the lavishness shown by another renowned Mexican miner of the eighteenth century, José de la Borda, who threw 400,000 pesos of the fortune he had made at Taxco into the building and decoration of the magnificent church of Santa Prisca in that town.[28] By contrast with these outlays, López's generosity to even his favored Franciscans seems to have been well controlled. And to the poor he was, it would seem, still less forthcoming. Arzáns, who in general admired López, and praised him, is on this point quite reserved, and even critical. After recounting López's gift to Viceroy Lemos for the layette of his newly born child, Arzáns goes on:

It is true that he had enough for any purpose, since God had given him so much; but that was no reason for him to lay out such a large sum of money to so little advantage. For there were so many poor people in the town with whom he could share what God had granted him, and in this respect others who were far less rich outdid him greatly in Potosí. I do not say that he did not give alms, but that for someone so wealthy, they were less than they should have been.[29]

Even López's funeral was less lavish than was to be expected of someone of his wealth, in Arzáns's view.[30] And so a certain parsimony begins to gather around the figure of the *maestre de campo*—a quality consistent, at least, with another of Arzáns's observations, to the effect that López, in contrast to many other wealthy figures in Potosí, did not squander his riches on vice;[31] and also with his remark that López was the only rich man known of in Potosí whose grandchildren had inherited his estate intact.[32]

Some tightness of fist, then, at least so far as spending unrelated to business or personal advancement went, marked the *maestre de campo.* This is far less than proof that he had a particularly active disposition to invest, but it does hint at a focussed mentality, an inclination to concentrate on what was productive; and these are traits consonant with, indeed conducive to, investment.

ENTERPRISE

Despite the disappointing lack of hard numbers, then, that would show beyond dispute such matters as López's profits, his absolute and comparative rates of investment, and the relationship between his mining and non-mining incomes, there can still be little doubt that in the context of seventeenth century Potosí he was a man of outstanding wealth, unusual powers of organization, and exceptional enterprise. Indeed, by almost any of the many definitions of the term, López must qualify as what is now called an entrepreneur. He was alert to business opportunities, willing to face future uncertainties in the conduct of his affairs, confident enough of his methods not to be deflected by the inherent riskiness of mining, quick to implement useful technological innovation, and, above all, an organizer, integrater, and a rationalizer of the productive processes that were necessary to find silver ore, extract it, and turn it into metal.[33] Some of these qualities he was undoubtedly born with, but others he learned, or at least brought to their full power, in Potosí. The capacity for learning that he displayed is, in fact, one of his most striking traits. He absorbed all aspects of the silver trade—mining, refining, finance, mintage— within a dozen years of his arrival in Potosí; spent the next decade, to the early seventies, improving on what he had learned; and then, for the remaining quarter-century of his life, reaped the benefits of those given, and self-generated, lessons. He became the master silver-maker of the Potosí district and, since that district was still the largest single source of silver in Spanish America, all indications are that López was the master maker of silver in the New World.

His failures in mining were few, and quickly overridden. Other failures were, apparently, if not forgotten, put behind him. The two that stand out are the fruitless effort to find the Gran Paitití and the denial of titles of nobility. It is hardly surprising that the quest for the Paitití came to nothing, because there was nothing to be found—or, at least, nothing on the scale that legend, and Antonio López following the legend, predicted. Incas may have gone in small numbers through the forests to the Parecis range, but no rich kingdom came of it. A lesser cause of failure, at least in the sense of López's having to spend excessively large sums before realizing that the venture was hopeless, was incompetent management of the expeditions by his nephew, don Benito Rivera y Quiroga, who, as a result of his failings, passed from being an apparently favored relative to one at odds with the *maestre de campo*. This was one case where López's preferred reliance on family members to attend to his interests did not yield the results that he hoped for. In the many other instances where he used relatives to run his concerns, success generally followed. Perhaps, apart from any possible personal incompetence, don Benito was simply operating too far away from Potosí for his uncle to be able to keep an eye on him and provide a general guidance for his efforts.

Titles of *adelantado*, count, and marquis eluded Antonio López for more complicated reasons. These combined some overreaching on his part, since full-blown seigneurial lords were not what the home government wanted to see appearing in the colonies, with, perhaps, a pervasive aversion to Galicians among many other Spaniards, and also an indefinite notion that López himself, socially speaking, was not altogether respectable. What is curious about this setback, and the same goes in some measure for the failure with the Great Paitití, is its rearward-looking, even medieval, frame of reference. In the case of titles, López, quite conscious that his lands stood astride the border between the civilized and the barbarian, wanted the medieval ranks of *adelantado* and march-lord (marquis) that were particularly appropriate to such a situation. His ambition to uncover the hidden kingdom of the Paitití also has its strong medieval resonances—knightly quests, domination and integration of the infidel, as in the Reconquest of Spain, and overt references to the conquest of Spanish America itself, a reaffirmation of medieval traditions that had occurred only a century or so before Antonio López was born. In fact, it was not only in connection with the Paitití that he compared himself (advantageously, of course) with the *conquistadores* of the previous century, but also in some of his mining ventures. In Porco, in particular, he saw himself as the emulator of the Pizarros, who had been among the first Spaniards to profit from those deposits. So, then, Antonio López was a man who faced both ways in time. Mostly he turned to the future, generally in the sense of being a businessman whose methods are recognizably those used by the most successful producers in the centuries between his time and ours, and specifically in his planning for his own, immediate ventures in mining and other enterprises. He was a man who strained against the trend of decline in seventeenth-century Charcas, looking and planning ahead, seeking to revive what was old and worn-out with his new combinations of techniques. But, from time to time, and especially during the 1670s, when his success in business and his confidence in himself stood higher than ever before, he could look behind, to aspirations and marks of success defined long in the past. Perhaps it was the seventeenth-century Spaniard in him making itself felt—a temporary atavism, as it were, in the flush of his material success, an introverted looking-back, of the sort so common among his contemporaries in the Peninsula, to earlier times when Spain was rising, glorious, just, and right. López's career in every other respect is a contradiction of such retrospection— an inclination that so fettered the Spain of his century. But even in him, perhaps, a few seeds of such attitudes remained, and in the heat of his success and satisfaction in the seventies, they could come to life. But they did not find fertile ground in him. Reverses crushed the sprigs of medieval hankerings, and Antonio López turned forward again, as his true character bid him do, to other purposes. If he could not have titles of high nobility, then he would work for the effective political domination of Charcas by covert means; if he could not conquer great and mysterious rulers, then domination of the sort that his experience showed he could accomplish would have to do. He would enlarge his estates, expand his agriculture, diversify his business, and plan to leave his physical mark on Potosí by driving the first galleries to pierce the Rich Hill through and through.

THE ENTREPRENEURIAL CONTEXT

If it is true that Antonio López fashioned a career that can properly be called entrepreneurial, then questions of the historical context in which he did so arise; for entrepreneurship is not a type of conduct that leaps to the mind of anyone contemplating the Spanish world of the seventeenth century. On the contrary, that century seems a time of many-sided decline, at least in Spain itself, and of clinging to traditional practices in business as in many other aspects of life, even when the old ways manifestly did nothing to improve the situation. So it is necessary to ask how a figure like Antonio López de Quiroga fits into, or could emerge from, the Hispanic world of his day.

Ther first broad observation that should be laid down is that those impressions of backwardness, conservatism, and enervation in Hispanic business belong more to Spain itself than to the colonies. In America, entrepreneurship was by no means rare in the seventeenth century, and had in reality been a marked feature of colonial life from the conquests onward. The conquerors themselves often enough demonstrated an avidity for profitable business, and a capacity for organizing it, that would merit the respect of any nineteenth- or twentieth-century industrial mogul. Cortés himself is a prime example in this line, with his enterprises for mining, farming, stock-raising, sugar production and marketing goods paid to him by the Indians in tribute.[34] In the decades after the conquests, a keen pursuit of productive enterprise is a feature of *encomenderos'* activities.[35] Officers of the royal administration were often, in the same period, among the most active of business people, using their extensive powers to promote their affairs. The *Licenciado* Lorenzo de Tejada, an early judge in the high court of New Spain, rapidly became after his arrival in Mexico in 1537 an energetic gatherer of land, a builder of irrigation works and flour mills, and a producer of livestock, wheat, grapes, and mulberries.[36] Don Antonio de Mendoza, first viceroy of New Spain, was himself no laggard in similar pursuits. Later on in the sixteenth century, at a time when the *encomenderos'* early position of privilege and *de facto* power in commanding Indian labor and control of land had been undermined by the crown, and when the freedom of governmental administrators to go into business had at least officially been restricted, other men had no difficulty in acting in entrepreneurial ways. Alonso de Villaseca, not an *encomendero*, but reputedly the richest man in Mexico by 1600, was a miner, farmer, and stockman on an impressive scale in the center and near-north of the colony. The Rivadeneiras, in the late sixteenth and early seventeenth centuries in Mexico, grew rich on the silver of Pachuca, and then became great landholders in the far south.[37] In Peru at the same period even certain Indian *curacas* come onto the business scene with all the appearances of entrepreneurship. The figure of don Diego Caqui, *curaca* of Tacna in southern Peru in the 1580s, has long been familiar to historians. He sold wine to Potosí that he had produced at a coastal vineyard planted with no fewer that 40,000 stems. These and other lands in his possession, together with three ships for coastal trading, gave him a net worth of 260,000 pesos.[38] Caqui was far from being the only native leader to trade agricultural produce to Potosí around the turn of the century.[39] Indeed, the examples of what by anyone's standards is clearly entrepreneurial behavior in the Spanish America of the sixteenth and seventeenth centuries go on and on .

It is clear enough that the colonies offered a more encouraging environment for entrepreneurship than the Peninsula for most of that time. In the first place, the abundance of labor (at least for the first few decades after the conquests) and of land, in combination with the authority that the colonizers could carry over from their military domination of the native populations, acted to stimulate entrepreneurial inclinations likely to have remained dormant or less fully expressed in the constrained conditions of the Peninsula. Then again, crossing the Atlantic produced some levelling of social distinctions and a consequent weakening of inhibitions on behavior. Just as commoners thought themselves improved in stature by virtue of completing the hazardous jour-

CONCLUSION

ney to the New World, and thereby more fitted to undertake large ventures than before, so men of established rank in Spain could benefit in the more flexible social conditions of America, if they were so inclined, from a relaxation of cultural restraints, well rooted at home, on nobles' participation in trade and production.

By the early seventeenth century, yet another stimulus to entrepreneurship may well have been appearing in the colonies, as the outcome of broad economic and political developments in the Spanish New World. Briefly put, the argument runs like this. Throughout the seventeenth century, and well on into the eighteenth for that matter, Spain's control, broadly defined, of her American territories weakened. Menacing political and military problems in Europe diverted the attention of the home government away from the Empire. Financial over-extension in the metropolis led to the sale of office, which allowed the growing population of American-born Spaniards access to ever higher levels of the imperial administrative system. Spanish merchants' attempts in the sixteenth century to keep prices for exported goods high in America by starving the markets led (along with a natural increase in artisan skills in the colonies as Indians were trained in Spanish crafts) to local production of what had previously been brought from Spain—and also to expansion of contraband trade. In economic matters, all this was conducive to increasing freedom of the market: taxes were less fully collected than before, ports whose trade was legally restricted (such as Buenos Aires) could not be effectively sealed, forbidden trade routes (such as the one between Mexico and Peru) remained open. If then, as one student of entrepreneurship maintains, one feature of a free market is freedom of entrepreneurial entry,[40] entrepreneurship would be expected at least to continue flourishing, or even to grow in vigor, as Spanish America moved farther into the seventeenth century.

One example that can be taken to support this prediction is that of the *peruleros*, those Lima-based traders who from the opening years of the century took advantage of Spain's decline in ability to enforce the laws regulating transatlantic trade to insert themselves into that commerce. The device they used was in essence a simple one. Instead of going, as before, to the fairs held at Porto Belo at the arrival of the trading fleets from Spain, there to buy merchandise from Spanish exporters, they now began to travel directly to Spain themselves. Once there, in Seville, they would buy the goods they were confident of selling in Peru directly from foreign suppliers. On the return voyage, they could simply by-pass the traditional fair, with its associated costs of transaction and delay, and move their cargos quickly and directly over the Isthmus of Panama to ships waiting in the Pacific. By avoiding the fair, they also were able to escape paying most of the customs charges normally levied on these transatlantic exchanges. Further, of course, they avoided paying for the profit that the Seville merchants had always sought to make on sales at Porto Belo. The result of this simplification of the processes of exchange—a true entrepreneurial rationalization if ever there was one—was to give the *peruleros* growing profits at the expense of the Seville traders' losses. And the end effect of that was to transfer control of the trade from Spain to Peru.[41]

Mining, too, always offered scope to entrepreneurial spirits in the Indies. Opportunities full of financial risk and uncertainty, chances to develop and implement potentially rewarding technical change, openings for investment, challenges of organization and integration—in fact, all the aspects of the entrepreneur's world that appear in Antonio López's career in Potosí—are to be found from the beginning in mining establishments across the colonies. And the entrepreneurs are to be found, also: Nicolás del Benino, and other builders of adits in the Hill of Potosí in the sixteenth century; or Diego de Ibarra, who besides being among the leading miners of Zacatecas before 1600, was also the owner of one of the largest estates in Mexico, and governor of New Biscay to boot.[47] And then, in the seventeenth century, there was another Zacatecan miner, Bartolomé Bravo de Acuña. He, like López, was an emigrant from Spain who crossed to America in the 1640s. In Zacatecas he indeed did in mining very much as López was to do in the sixties in las Amoladeras in Potosí, taking four abandoned and flooded mines, which he bought cheap, and interconnecting

them so that they could be drained, and then worked more efficiently than before. This was the start of increasing integration
of workings in Zacatecas that continued, with profitable results,
until the early nineteenth century.[43]

But López's career, in its range of activities and its achievements,
stands out above that of any known earlier figure in mining.
Though much diminished by the time of his death from its stature of a century earlier, the Potosí district was still the leading
source of silver in Spanish America. And no one before him had
dominated the Potosí district as he did.

Is there anything further, then, that can be added to account
for this career—besides López's remarkable personal qualities and
the general conditions favoring entrepreneurship in the Indies of
seventeenth century that have already been described? One further possibility may indeed have some application to this case.

This is the suggestion put forward in recent years that, within
the gloom that covers Spanish economic life during the final
decades of the seventeenth century—that generally murky era associated with Charles II, the last of the Habsburg kings of Spain—
some signs of recovery and material advance can nevertheless be
glimpsed. One of these is a nascent trend precisely toward entrepreneurship. An "ineradicably Catholic capitalist ethic" is said
to have appeared in the country.[44] Evidence for this, apart from
specific instances that can be found, is the publication of royal
pragmatic in 1682 that indicates that ancient cultural restraints
on the pursuit of business and industry by nobles had begun to
weaken.[45] The decree attempted to reassure nobles that their
aristocratic status would not be prejudiced if they engaged in
such pursuits. It stated, in summary, that "to maintain . . . manufactures . . . is not contrary to the quality, immunities and
prerogatives of nobility . . . so long as those who . . . maintain
manufactures do not labor with their own persons but through
their servants and officials."[46] The qualification made in the second part of the statement would seem to rob it of some of its
force for change (especially since members of the lower nobility
had already been known to engage on occasion in highly lucra-
tive commerce, such as the Indies trade in the sixteenth centu-

ry); but nevertheless it may be true that the decree is important precisely because it signalled that the longstanding antipathy among the nobility to business was already relaxing. This change may have been a response to the extreme prostration of the nobility, especially of the *hidalgos*, or petty nobility, whose numbers were greatest in northern Spain.[47] Since Antonio López came from a northern family, although he left Spain a good forty years before the 1682 pragmatic was issued, he may fall within some such pattern.

It is possible, then, that Antonio López de Quiroga's remarkable activities in and around Potosí were the outcome of three influences: traits within himself that fitted him for, and drove him into, a career of that sort; an openness in the economic system operating in Spanish South America by the time he came on the scene, favorable to entrepreneurial effort, and reinforcing the tradition of entrepreneurship already clear in the Indies; and some broad trend toward a new activism by the upper end of Spanish society in matters of business and production. Of this trend he may have been a spectacular exemplar.

From his time onward, expanding entrepreneurship certainly is not hard to find in some parts of Spanish America. It has long been known that mid- and late-eighteenth century New Spain was rich in great mining venturers, José de la Borda being just one among many.[48] Zacatecas, at least, also possessed mining entrepreneurs in the first half of that century, men following the path opened by Bravo de Acuña.[49] Entrepreneurship in Mexican textile making before 1750, as well as commercial and manufacturing entrepreneurship in Mexico City after then, have recently been identified and examined.[50] But where Mexico was rich in enterprise, Peru and the central Andes in general may not have done so well. Antonio López may have had an emulator in Potosí in the early eighteenth-century miner José de Quirós, another product of Galicia, and another man to benefit from the wealth of the Amoladeras sector of the Rich Hill.[51] But in general, Andean mining after 1700, in Potosí and elsewhere, seems to have been a fragmented affair, without dominant, unifying figures, and in fact characterized by a multiplicity of lesser operators.[52] Just why this

CONCLUSION

divergence should have developed between the central Andes and New Spain remains to be seen.[53] The effect of its existence, though, and of the apparent absence of great entrepreneurial figures in Peru and Charcas, the central Andean colonies of Spain, after 1700 is to leave the *maestre de campo* Antonio López de Quiroga towering over his historical landscape in magisterial solitude.

Abbreviations and Conventions Used

AAGN   Argentina, Archivo General de la Nación
AGI   Archivo General de Indias
BAE   Biblioteca de Autores Españoles
BAN   Bolivia, Archivo Nacional
CPLA   Cabildo de Potosí, Libros de Acuerdos
CR   Cajas Reales
DII   *Coleccíon de documentos inéditos relativos al descubri-miento, conquista y organizacíon de las antiguas posesiones españolas de América y Oceanía, sacadas de los archivos del Reino y muy especialmente del de Indias.* 42 vols. Madrid, 1864–1884.
EN   Escrituras Notariales
HAHR   *Hispanic American Historical Review*
JHR   Information gathered by Judith Hope Reynolds
PCM   Potosí, Casa Nacional de Moneda
SRAH   Spain, Real Academia de la Historia

References to laws from the *Recopilacíon de leyes de los Reinos de las Indias* of 1681 are given in the order: book, title, law. Thus, *Recopilacíon* 1, 6, 26 signifies Book 1, Title 6, Law 26.

Cash amounts are stated in *pesos de a ocho* of 272 *maravedís*. These coins were sometimes known as *patacones* or *pesos corrientes*. Each was the equivalent of 8 reales of 34 *maravedís*.

NOTES

PROLOGUE

1. "El dicho verdugo lo arrojó de la escalera y se subió sobre los hombros de dicho negro, y le pisó muchas veces hasta que pareció estar naturalmente muerto y quedó colgado del pescuezo de la dicha horca, y en esta forma se hizo justicia del dicho negro Juan Luis Osorio. . . ." For this case, and the description of the execution, see PCM, Archivo de la Casa Real de Moneda ("las cajas de Don Armando"), caja 6, legajo 71: "1656. Criminal. Que de oficio de la Real Justicia se sigue contra Juan Luis Osorio, negro esclavo de Antonio López de Quiroga, sobre la resistencia que hizo al dicho su amo y a los señores alcaldes Jueces. . . ." The description of the execution is on f. 68v.

2. "Aquel nuevo mundo"—so described by Antonio López and don Benito de Rivera y Quiroga to an unidentified "excelentísimo señor" in Spain (asking him to recommend their effort to the queen) in a letter dated Potosí, March 14, 1670 (AGI Charcas 23).

3. The scene is described by witnesses to an "Información" about the services of Antonio López bearing the final date of Potosí, November 22, 1690 (AGI Charcas 128). See f. 35, testimony of don Baltasar de Guzmán, Potosí, December 2, 1689; f. 47v., testimony of Ignacio de la Cueva, Potosí, December 6, 1689; and f. 71v., testimony of Captain don Gerónimo Antonio de Taboada y Encalada, Potosí, December 29, 1689. The enterprise of the Gran Paitití, begun in 1669, is discussed below in ch. 3.

4. The connection with workings on the Veta Rica was made at 2 A.M. on August 28, 1677. See the letter of don Cristóbal de Quiroga y Osorio, *corregidor* (local governor) of the province of los Lipes, to the *vicario* don Francisco de Bóveda y Saravia (brother-in-law of Antonio

López) in Potosí, dated los Lipes, August 29, 1677 (AGI Charcas 23). López's undertakings at San Antonio are described more fully in chapter 3 below.

5. The final outcome of Azañón's complaints against Antonio López and don Diego Reinoso is not known. Both his version of the matter, and Reinoso's, reached the Council of the Indies in 1693. In August of 1694 the Council decided to send the case back to the Audiencia of La Plata. See the "Extracto de lo que ha pasado en los particulares de Don Gregorio Azañón, que fue corregidor del asiento de los Lipes en el Reino del Perú," seen in the Council on August 7, 1694 (AGI Charcas 61). Azañón's letter to the Council laying out his side of the case is dated La Plata, October 18, 1692 (AGI Charcas 62). Another detailed, if dramatizing, account of what happened in San Antonio, dated los Lipes, March 28, 1692 was sent by a number of miners and inhabitants of the town to an unidentified "Muy poderoso señor." The list of signatories begins with Pedro López Tufino and Miguel Grimaldos de Ayala (AGI Charcas 62).

CHAPTER I

1. The first page of a possibly holograph letter of Antonio López to the king, marked on the back "Potosí. A su mgd. 1674. Antonio López de Quiroga. 31 de diciembre." (AGI Charcas 128).

2. "Amistad y buena obra." (PCM EN 112, ff. 590–91v., Potosí, April 24, 1649.) Bartolomé Arzáns de Orsúa y Vela, the chronicler of Potosí, whose prolixity is usually greater than his chronological precision, also states that López arrived in Potosí in 1648 "in the prime of his life [and] in search of riches. . . ." See his *Historia de la Villa Imperial de Potosí* (ed. Lewis Hanke and Gunnar Mendoza L., 3 vols., Providence: Brown University Press, 1965), vol. 2, p. 395.

3. López's burial in the Franciscan monastery of Potosí, on January 24, 1699, is recorded in the Archivo de la Catedral de Potosí, Libro de entierros No. 2 (1680–1701), f. 196v. The author is indebted to the late don Mario Chacón Torres for a copy of the entry.

4. In 1678, in Potosí, López testified to the good qualities of one Captain Agustín de Rojas, a vecino of Seville and a merchant who had recently died in poverty in Potosí. López recalled having known Rojas in Seville as a shipper, and as a large importer in Lima in 1643. (PCM EN 131, ff. 717v.–18v., testimony given on October 29, 1678). See also n. 24 below.

183

5. Arzáns, *Historia*, vol. 2, p. 398. In his *Anales de la Villa Imperial de Potosí*, a summary of the *Historia*, Arzáns further notes of López that "in the end he was of such a great age that it was necessary to sustain him with milk from the breasts of women, allowing him to nurse." See the *Historia*, vol. 2, p. 398, n. 5 by Gunnar Mendoza L. To put this observation in context, it is worth recalling that sixty was an advanced age for Europeans of early modern times, even for people of wealth.

6. The birthplace is stated in the record of López's burial in the Cathedral of Potosí. See n. 3, above.

7. *Enciclopedia universal ilustrada europeo-americana* (Madrid: Espasa-Calpe, 1908–), vol. XLVIII, p. 1,432.

8. Alberto y Arturo García Carraffa, *Enciclopedia heráldica y genealógica hispano-americana* (Madrid 1952–63), vol. 76, p. 78.

9. Licenciado Luis Molina, *Descripción del Reino de Galicia y de las cosas notables de él, con las armas y blasones de los linajes de Galicia, de donde proceden señaladas Casas de Castilla* (Mondoñedo 1550)—cited in Javier Ruiz Almansa, *La población de Galicia, 1500–1945* (Madrid: Consejo Superior de Investigaciones Científicas, 1948), vol. 1, pp. 51, 116.

10. García Carraffa, *Enciclopedia*, vol. 76, p. 79.

11. The surname López, indeed, seems to have been brought into the Quiroga line by the marriage of Juan López de Mosquera, lord of the house of Espasantes, situated near Monforte, to Violante de Taboada y Quiroga in the late sixteenth century. See García Carraffa, *Enciclopedia*, vol. 76, pp. 82–88. Genealogical information on López's father and uncles is taken from Julián de Paredes's dedication, addressed to Antonio López de Quiroga, to *Nobiliario, armas, y triunfos de Galicia, hechos heroicos de sus hijos, y elogios de su nobleza, y de la mayor de España, y Europa, compuesto por el Padre Maestro Fray Felipe de la Gándara, de la Orden de San Agustín, Coronista General de los Reinos de León y Galicia* (Madrid 1677), f. 3. Paredes was the printer of this work, which appeared posthumously, Gándara having died after he had sent the manuscript to the press.

12. Ruiz Almansa, *La población*, p. 116

13. The reasons for this omission can only be guessed. A sole manuscript has appeared in which López includes his mother's family name. This is an "Expediente promovido [por el] maestre de campo Antonio López de Quiroga, minero [y] azoguero del Potosí, [sobre] que se le eximiese de las alcabalas y unión de armas en sus haciendas. Años de 1674 y 1697" (AGI Charcas 128). In this is a document, apparently

of 1670–72, signed by Antonio López, and marked "No. 2," on the verso of which is written "El capitán Antonio López de Quiroga, vecino de la Villa Imperial de Potosí. Excelentísimo senor." López begins this by stating that he descends from the three Galician *casas solariegas* of Quiroga, Reimóndez, and Rivera, "all three of noble lineage," and all having knights of the military orders in their ranks. Rivera, to be sure, was a highly distinguished name. The family had ancient origins in Galicia, and had extended over much of Spain, gathering many titles of nobility. The connection with the Quirogas seems to have developed especially in Medina del Campo, near Valladolid, where in the late sixteenth century one Pedro de Rivera y Quintanilla married doña María de Quiroga y Zúñiga, sister of Cardinal Gaspar Rodríguez de Quiroga. A prominent figure in the later life of Antonio López de Quiroga was his nephew, don Benito Rivera y Quiroga, who was perhaps descended from this union. (See the discussion in the text, below, of the exploration of the Gran Paitití.) For the Riveras, see García Carraffa, *Enciclopedia*, vol. 78, p. 243, and pp. 218–251 in general.

It is also worth recording, since the connection between Antonio López and the Somoza family of Galicia has some later significance, that he did once add this surname to his usual López de Quiroga. This was in his first petition for the title of Count, made in 1674. (See AGI Charcas 128, ms marked on verso "Potosí. A su mgd. 1674. Antonio López de Quiroga. 31 de diciembre.") The standing of the Somozas was on a level with that of the Quirogas and Reimóndez's, although the connection between Somozas and Quirogas is not clear. As later discussion here shows, however, one don Juan Somoza Losada y Quiroga was present in Peru, as governor of Santa Cruz de la Sierra, among other things, from the late 1630s onward.

14. García Carraffa, *Enciclopedia*, vol. 76, p. 83. Don Rodrigo de Quiroga was highly recommended to the king in 1564 by the then governor of Peru, Lope García de Castro, who described him as a "very rich and important man in Chile . . . and very well liked." The king could do no better than to appoint Quiroga to govern Chile. See García de Castro to king, AGI Lima 92, Los Reyes, November 20, 1564.

15. "The famous voyage of Sir Francis Drake into the South Sea, and there hence about the whole globe of the earth, begun in the year of our Lord, 1577," in Richard Hakluyt, *Voyages and discoveries. The principal navigations, voyages, traffiques and discoveries of the English nation*, ed. Jack Beeching (Harmondsworth 1972), pp. 171–88.

16. Emilio González López, *La Galicia de los Austrias*. Tomo 1.

1506–1598 (La Coruña: Fundación "Pedro Barrie de la Maza, Conde de Fenosa," 1980), p. 376.

17. García Carraffa, *Enciclopedia*, vol. 51, pp. 220–22. The fourth and fifth lines of the Losada family were especially closely entwined with the Quirogas. The degree of interconnection reached such a point in the sixteenth century that, in one case, the same man seems to be identifiable by two sets of surnames, one from each family: don García de Losada y Ribadeneyra is apparently the same person as don García de Quiroga. (See González López, *La Galicia*, vol. 1, p. 375. This inconsistency is not as odd as it might seem to readers familiar with the customs of the English-speaking world. The Spanish use of surnames is far more flexible, especially in past centuries, than the English. Out of personal preference, and also to suit the needs of the moment, a person might call himself, at one time or another, by a changing variety of his forebears' names.) Don García was father to don Juan de Losada y Correa (also known as don Juan de Losada y Quiroga), who was the first Losada to make an appearance in Chile, arriving there in 1557. Don Juan fought against the Araucanians with don Rodrigo López de Quiroga, and occupied posts in the local town governments of los Confines and Santiago —finally dying aboard ship on his way back to Spain in 1575. See González López, ibid.

It is also worth noting that, apparently early in the seventeenth century, the Somoza family became connected with the fourth branch of the Losadas, by the marriage of Captain Diego de Losada y Taboada to doña Isabel López de Somoza (ibid., p. 221, n. 1, III). While it is not possible to establish any definite link between Antonio López de Quiroga and these people, some connection is suggested by the fact that Antonio López, as already remarked, once gave his full name as Antonio López de Quiroga y Somoza.

18. García Carraffa, *Enciclopedia*, vol. 51, p. 223, para. III.

19. The will was of Domingo Santos, *morador* (dweller) and *vecino* (householder) of Potosí. (PCM EN 53, f. 2,798v., Potosí, September 3, 1620). López and Pedro de Tapia Cevallos received a power of attorney to look for mines from Agustín Ruiz de Porres and Pedro López Pallares, notary public, on September 9, 1620. (PCM EN 53, f. 2,683–83v.)

20. What Antonio López guaranteed was that the corregidor, don Pedro de Yebra Pimentel, would pay what he still owed for his *media anata*—a tax on the salary paid to a crown official during his first year of office. (PCM CR 264, f. 114v.). What also catches the attention here is that Pimentel is a Galician surname. It would be naturally quite expected

that Galicians, like other Spanish regional groups in America, would support each other with guarantees and the like.

In the early 1630s another Quiroga was active in and around Potosí. This was Lorenzo de Quiroga, who in 1633 appears as a small scale silver miner in the town, and in 1635 as the owner of a small grain and stockraising farm in the Canasmoro valley, some 130 miles southeast of Potosí. In 1633 he received an allotment of twenty *mitayos*, or forced Indian laborers, in the *repartimiento*, or distribution, of such workers made by don Juan de Carvajal y Sande. Quiroga is classed here among the *soldados mineros* of Potosí, which means that he was probably more of a prospector than an established silver producer. *Soldado* was a term dating from the early days of the central Andean mining towns, and used to describe men who went hunting fortunes by looking for new ore deposits. Some of the original *soldados* may actually have been soldiers. (AAGN Sala 13, cuerpo 23, 10–2. "Apuntamiento general hecho por el Dr. don Dionisio Pérez Manrique . . . presidente de la Real Audiencia de La Plata, de los indios efectivos que acuden a la mita del cerro rico de Potosí por el repartimiento hecho por el señor don Juan de Carvajal y Sande. . . ." The cover date is November 7, 1643, but Carvajal's distribution had been made ten years earlier.) Quiroga's piece of land was a *chácara de pan llevar*, or grain farm, on which he also raised a few cattle and sheep. He gave half of it as a dowry to a young woman named Juana de Caviera, who he said was a *niña* he had raised in his house. (PCM EN 88, f. 575, Potosí, January 29, 1635). This Quiroga's connection with Antonio López is unprovable, but again the common interest in mining is impossible to ignore.

21. On December 31, 1635 Somoza Losada y Quiroga gave a power of attorney to one Francisco Ortiz to run his mines in Carangas. (BAN Minas t. 144 (Minas catalog No. 852), ff. 1,341v.–42).

22. On the other hand, don Juan most certainly had a younger brother, aged fourteen in 1637, named don Andrés de Somoza y Quiroga, who accompanied him to Peru as a servant (*criado*) when he went to take up his governorship. (AGI Contratación 5,419, ramo 23, "Don Juan de Losada y Quiroga.")

23. Ernst Schäfer, *El Consejo Real y Supremo de las Indias. Su historia, organización y labor administrativa hasta la terminación de la Casa de Austria*. Tomo II. *La labor del Consejo de Indias en la administración colonial* (Seville 1947), p. 553.

24. See, for Antonio López's powers from don Juan de Somoza Losada y Quiroga, the "Información en conformidad de una cédula real de su

magestad . . . de los méritos y servicios del gobernador de Santa Cruz y corregidor de las fronteras de Tomina, y justicia mayor de Pomabamba, que es el general don Juan de Somoza y Losada. . . ." (AGI Charcas 57, final date at La Plata, March 10, 1643.) For don Juan's origins, see a power of attorney issued by him, November 26, 1655, at Espíritu Santo de Carangas, north of Potosí, to his uncle, don Francisco de Quiroga y Taboada, his uncle the *abad* Alonso López Vizcaíno, and his brother, don Francisco de Somoza y Quiroga, *clérigo presbítero*, to administer the family estate at Laxosa, consisting of a *palacio viejo y mayorazgo de la casa solariega.* (BAN Minas t. 144 (Minas catalog No. 852), f. 1,319–19v.)

25. *Historia*, vol. 2, p. 395.

26. "Generosa y originaria casa de este reino . . . ," according to Julián de Paredes in his dedication, addressed to Antonio López de Quiroga, to Gándara's *Nobiliario*, f. 3.

27. Don Francisco de Bóveda y Saravia describes his parents as *personas nobles* in a report on his services presented to the Audiencia of La Plata in 1674. See the *parecer*, or opinion, of the high court on this report, dated La Plata, November 26, 1674, in BAN, Audiencia de Charcas [i.e. La Plata], Libros de acuerdos, vol. 13, f. 232. This document also confirms the marriage of Lorenzo de Bóveda with doña Ana María Bravo de Saravia. Doña Felipa's marriage to Antonio López de Quiroga is confirmed by a reference in PCM EN 128, f. 245, in 1676.

For brief information on the Bóveda family's past, see García Carraffa, *Enciclopedia*, vol. 18, p. 184. The Bravo de Saravias had their origins in the Castilian city of Soria; but the family had become strongly established and prominent in Chile during the sixteenth century. (Ibid., vol. 19, pp. 24–26.) It is therefore possible that doña Ana María, Antonio López's new mother-in-law, had some connection with earlier Quirogas in that part of South America, though this remains to be shown.

28. *Fletamento* by Juan Gago, "dueño de recua de mulas de la carrera de la frontera de Santa Cruz de la Sierra," for Lorenzo de Bóveda, *vecino* and *mercader* of Potosí, Potosí, January 13, 1640 (PCM EN 104, ff. 321–22v.).

29. The loans are recorded for October 19–20, 1648, in PCM EN 113, ff. 1,949v.–50v., 1,983–83v., and 2,067 67v. Prices of houses and llamas are taken from various sale contracts recorded in notarial registers for 1648 and 1649 (in PCM EN 112, 114). The characterization of Bóveda as a *mercader grueso* is in "Demanda. El licenciado Gaspar González

Pabón contra el 24 don Gerónimo Julián Mejía, sobre el interés y daños de 4,000 pesos corrientes," no place or date, but mainly Potosí, 1647–52 (AGI Escribanía de Cámara 865C, "pieza 5a," f. 112).

30. A power of attorney (poder) granted by Antonio López, as executor of Bóveda's estate, to one Joseph García del Campillo instructs him to collect 3,300 pesos lent by Bóveda in 1650 to one Francisco de Monroy in the mining center of Titiri. While this loan could have been used for non-mining purposes there, the likelihood is that it was made for some purpose connected with producing silver (PCM EN 116, ff. 197v.–98).

31. Arzáns, Historia, vol. 2, p. 39. The power of attorney referred to in the preceding note defines López as executor, and also mentions the date of Bóveda's final will as May 4, 1652.

32. Between 1593 and 1597, the average of the Spanish crown's total expenditures, on all business of state in Europe, was 12,892,690 ducats a year. (I am indebted to Professor J. H. Elliott for this figure, which comes from a report discussed in the royal Junta de Hacienda on January 26, 1606. See Archivo General de Simancas, Cámara de Castilla, legajo 2,794, pieza 1, f. 692.) Potosí's registered (i.e. taxed) silver production in 1592 of 887,000 marks, if converted into ducats (taking the conventional valuation of the mark of fine silver at 2,380 maravedís, and that of the ducat at 375 maravedís), amounts to 5,629,493 ducats, or 43.66 percent of that average crown spending. The average annual tax gathered on Potosí's silver production—the royalty of a fifth, and other charges amounting to 1.5 percent—over the years 1592–96 was the equivalent of 1,127,235 ducats, or 8.74 percent of royal expenditure each year from 1593 to 1597.

For silver production in Potosí in the sixteenth and seventeenth centuries, see Peter Bakewell, "Registered silver production in the Potosí district, 1550–1735," Jahrbuch für Geschichte von Staat, Wirtschaft und Gesellschaft Lateinamerikas, 12 (1975), pp. 67–103.

33. For labor and refining techniques at Potosí, see Peter Bakewell, Miners of the red mountain. Indian labor in Potosí, 1545–1650, (Albuquerque: University of New Mexico Press, 1984), especially chs. 1 and 3; also Peter Bakewell, "Technological change in Potosí: The silver boom of the 1570s," Jahrbuch für Geschichte von Staat, Wirtschaft und Gesellschaft Lateinamerikas, 14 (1977), pp. 60–77. See also Jeffrey A. Cole, The Potosí mita, 1573–1700. Compulsory Indian labor in the Andes (Stanford: Stanford University Press, 1985), ch. 1.

34. For the notion that high-altitude adaptation among the native Andean population is an evolutionary acquisition, genetically transmit-

ted, see the classic work on the subject by Carlos Monge: *Acclimatization in the Andes. Historical confirmations of "Climatic Aggression" in the development of Andean man* (Baltimore: the Johns Hopkins Press, 1948), pp. 30, 38. A more recent view is that adaptation to altitude, at least as it appears in enlarged lung volume, is more a matter of development within the individual's lifetime than an inherited trait. For maximum increase in lung capacity, however, the individual must be exposed to altitude in early childhood. Adults moving from low to high terrain do not exhibit as much increase as children moved when very young. See A. R. Frisancho, "Human growth and development among high-altitude populations," in P. T. Baker (ed.), *The biology of high-altitude peoples* (Cambridge: Cambridge University Press, 1978), pp. 133–34, 163–66.

35. Around 1600, the total native labor force working directly in mining and refining in Potosí was some 10,000 men. Of these, roughly 5,500 were wage workers, or *mingas*. The other 4,500 were draft laborers, or *mitayos*. The number in the draft was about what Toledo's arrangements had provided for; but this was achieved only by making the draftees work more intensively than he had ordered. His plan had been that some 13,500 adult Indian men should form the draft force in Potosí each year, but that each man should work for only four months out of twelve. So the number working on any particular day would be 4,500. With declining numbers of *mitayos* coming to Potosí each year, that figure could be, and was, maintained only by obliging individuals to extend their period of labor beyond four months. By 1600, *mitayos* may have been working, on average, for six months out of their allotted twelve in Potosí. For numbers of laborers, Bakewell, *Miners*, pp. 127–28; and for wage labor in general, ch. 4, passim. See also Cole, *The Potosí mita*.

36. For wage rates, Bakewell, *Miners*, pp. 101, 125. The *minga* rate for ore cutting, around 1600, was 56–72 reales for a working week of five days; while the *mitayo*'s rate, at the same time, was 20 reales a week. By the 1630s, the upper limit for *minga* ore cutters seems to have been 60 reales; and the *mita* rate was the same as before.

37. For detail on these improvements, see the exhaustive treatment by Modesto Bargalló, *La amalgamación de los minerales de plata en Hispanoamérica colonial* (Mexico City: Compañía Fundidora de Fierro y Acero de Monterrey, 1969). A general description of the refining process with mercury is in Peter Bakewell, "Mining in colonial Spanish America," (*The Cambridge History of Latin America*. Vol. 2. *Colo-*

nial Latin America), ch. 4, (Cambridge: Cambridge University Press, 1984), p. 113 ff.

38. Bakewell, *Miners*, p. 127.

39. William E. Rudolph, "The lakes of Potosí," *The Geographical Review*, 26:4, (New York, October 1936) pp. 529–54.

40. In fact, if price is any guide to the availability of goods (as it should be, prices rising as demand exceeds supply, and falling as the reverse happens), supplies of many items in Potosí appear to have been more plentiful after 1600 than before. Prices of nearly all items for which useful records exist fall, or at least are steady, in the first half of the seventeenth century. For example, the price of a llama went down from between 6 and 9 pesos in the first fifteen years of the century, to 4 to 5 pesos between 1635 and 1649. Mercury, an essential and costly item for refiners, fell steadily in price from 85 pesos the *quintal* (100 lbs.) in the 1590s, to 58 pesos between 1645 and 1655. And various other goods of direct or indirect concern to silver producers—bar iron for general purposes, iron stamp shoes for the crushing mills, the coca leaves chewed by native laborers, the wine consumed mainly by Spaniards, wheat flour— all these show declining or, in one or two cases, stable prices between 1600 and 1650. Falling prices, of course, broadly meant falling costs for the producers of silver; or, to put it another way, the value of the silver produced in Potosí, at least when measured against goods sold in the local economy, tended to rise in the new century. This was a further advantage available to miners after 1600, but evidently inadequate to offset the rising costs of first, ore extraction, and, secondarily, labor. For prices of llamas, see Bakewell, *Miners*, Appendix 2, p. 194. Mercury supply was a monopoly of the royal treasury. Prices given here are from PCM Cajas Reales 503, f. 234–234v. Prices for the other items mentioned here are from various notarial registers of Potosí for 1587, 1589, 1594, 1599, 1604, 1609, 1614, 1620–21 (January-June), 1625, 1630, 1635–36 (January-May), 1640, 1645, and 1649. The individual references are too numerous to list.

41. Juan de Matienzo to the king, La Plata, October 20, 1561. Transcribed in Roberto Levillier, *La Audiencia de Charcas. Correspondencia de presidentes y oidores* (3 vols., Madrid 1918–22), vol. 1, p. 57.

42. Matienzo to king, Potosí, December 23, 1577 (Levillier, *Charcas*, vol. 1, pp. 455–56).

43. Alonso Mesía to the Viceroy don Luis de Velasco (n.d., but since the writer refers to a royal cédula of November 1601, and Velasco left office late in 1603, the report is clearly from 1602 or 1603), "Sobre las

cédulas del servicio personal de los indios,"(*DII*, vol. 6, pp. 146–47). Mesía notes that some people thought there were as many as 80,000 Indian men in Potosí at that time.

44. Don Pedro de Lodeña, *corregidor* of Potosí, to king, Potosí, April 9, 1603, para. 4 (AGI Charcas 46. JHR). *Chuño* is a form of dried potato still much used in the central Andes. *Quinoa* is a native Andean grain, rich in protein.

45. A reported, but unlocated, census of 1610 puts the total at 160,000: 76,000 Indians, 3,000 Spaniards born in Potosí, 35,000 Spaniards born in other parts of the Indies, 40,000 Spaniards from Spain, and 6,000 Blacks, mulattos, and *zambos* (children of Blacks and Indians). See Arzáns, *Historia*, vol. 1, p. 286. In 1600, for comparison, the populations of, respectively, Amsterdam, London, Seville, and Venice, were about 80,000, 130,000, 150,000, and 150,000. See Henry Kamen, *The iron century. Social change in Europe. 1550–1660* (London 1971), p. 21.

46. The most complete series of population estimates for Potosí is that provided by Hanke and Mendoza in their introduction to Arzáns's *Historia* (vol. 1, pp. xix–xx).

47. Arzáns, *Historia*, vol. 1, p. xviii, and vol. 2, p. 390.

48. This is perhaps the predominant feeling left in the mind by a reading of Arzáns' *Historia*. For a selection of examples, see Robert C. Padden's edited excerpts from Arzáns: *Tales of Potosí. Bartolomé Arzáns de Orsúa y Vela*, translated from the Spanish by Frances M. López-Morillas (Providence: Brown University Press, 1975).

49. "Historia de Huérfano [sic] por Andrés de León, vecino de la ínclita y nobilísima ciudad de Granada. Describe en ella muchas ciudades de las Indias. . . ." (SRAH, Colección Muñoz, ms A/70, ff. 180–203), f. 197v. This document is undated. A reference on f. 197 to 1612 places it after then; and allusions in it to factional strife between different groups of Spàniards in Potosí, which grew serious in the early 1620s, suggest that the author was in the town by then. (See, for this conflict, subsequent discussion in this chapter). A *real* was a silver coin weighing an eighth of an ounce, and valued at 34 *maravedís*. Eight *reales* made up one *peso*.

50. Ibid., ff. 197v.–98v.

51. Ibid., f. 200v.

52. "Recibo de dote" of doña Lorenza Quiroga Bóveda y Saravia, Potosí, October 21, 1676 (PCM EN 128, ff. 245–250v.), ff. 246v., 248v.

53. The first comment is by the *oidor* Juan López de Cepeda, cited in Arzáns, vol. 1, p. xxx. (For the full context, see Levillier, *Audiencia*,

vol. 2, pp. 137–38). For Arzáns's own observation, see the *Historia*, vol. 1, p. xxx.

54. *Historia*, vol. 2, p. 409. For a clear example of Arzáns's guilty (but only just so) delight in tales of violence, see the *Historia*, vol. 2, p. 149 ff., Capítulo VI "En que se cuentan los hechos de dos doncellas nobles naturales de esta Villa. . . ." (also in Padden, *Tales*, p. 58 ff.).

55. "Historia de huérfano . . ." (SRAH, Colección Muñoz, ms A/70), f. 197.

56. Ibid., f. 196v. This commentator estimated the number of hopeful idlers ("gente vaga") at a thousand.

57. Esquilache to king, "Gobierno, Duplicado, No. X," Lima, April 6, 1617 (AGI Lima 37).

58. Alberto Crespo Rodas, *La guerra entre vicuñas y vascongados. Potosí 1622–1625* (2d. ed., La Paz: Colección Popular, 1969), p. 164. This is the best account of the conflict. Also most useful is Gunnar Mendoza L., *Guerra civil entre vascongados y otras naciones de Potosí. Documentos del Archivo Nacional de Bolivia* (Potosí: Cuadernos de la Colección de la Cultura Boliviana, 1954). This is mainly a calendar of manuscripts in the Bolivian national archive on the topic.

59. Crespo, *Guerra*, p. 193.

60. Mendoza, *Guerra*, p. 29.

61. Ibid., p. 31.

62. An anonymous, undated description of Potosí in AGI Charcas 134 (first ms in the section of this legajo marked "sin fecha"), f. 2. The wording corresponds closely to that of the first part of the "Descripción de la Villa y minas de Potosí. Año de 1603" published by Marcos Jiménez de la Espada in *Relaciones geográficas de Indias—Perú*, vol. 1 (BAE, vol. 183, Madrid 1965), pp. 372–78; though "ichu" in the manuscript version appears as "paja" in Jiménez de la Espada's. *Ichu* is a rough grass that grows at great heights in the Andes, up to well beyond the tree line. Its use for roofing, fuel, and llama fodder goes back into the distant Andean past.

63. Ibid, f. 2v.

64. Lease by Pedro Venegas to the *Licenciado* Alonso Cabezas, of *altos de casas* above the store occupied by Gerónimo de Albadán, Potosí, November 27, 1614 (PCM EN 47B, ff. 2,593v.–94).

65. Information about the *casas de cabildo* is from PCM CR 207, ff. 1–60—a description of the *propios*, or sources of income, of the town council between 1625 and 1668 .

66. The going price around the beginning of the seventeenth century

was 5,000 pesos ensayados, or about 8,300 pesos de a ocho, for an aldermanship; though the price ranged up to twice that. See "Oficios que a pedimento del Dr. don Gerónimo de Tovar y Montalvo, fiscal de su magestad, se han vendido por su real hacienda" (AGI Charcas 17), no place or date, but c. 1600.

67. The Viceroy Marquis of Montesclaros to the king, Lima, November 20, 1609, reporting on the sale of this office to Hernando Ortiz de Vargas in Potosí (AGI Lima 35).

68. The *alferezazgo real* went to Diego Ortiz de Ortega in Potosí, around 1600, for 25,010 pesos ensayados, or some 41,375 pesos de a ocho. See "Oficios que a pedimento del Dr. don Gerónimo de Tovar y Montalvo . . ." (AGI Charcas 17).

69. "Gastos extraordinarios" of the town council of Potosí, c. 1625 (PCM CR 20, ff. 30–31v.).

70. According to Spanish law from the thirteenth century onwards, the subsoil was the property of the crown; and access to it was granted only in return for a royalty payment on any metal produced. See Walter Howe, *The Mining Guild of New Spain and its Tribunal General, 1770–1821* (Cambridge: Harvard University Press, 1949), pp. 2–3. In Potosí and its district, the royalty due was a fifth—until 1736, when it was cut to a tenth.

71. The full name of the chest was the *caja de las tres llaves*. All branches of the treasury possessed one, or more, according to need.

72. The leasing of the gaming room, where *tablas*, a game apparently similar to checkers, was played, is recorded in PCM EN 76A, ff. 4,173v.–75. The rent was 360 pesos for one year, and included a bedroom, a second room off the patio of the house, and a closet, as well as three large gaming tables, two settles, nine benches of various sizes, ten boards, nine yards of chequered baize, six pottery candlesticks, two pairs of scissors for trimming wicks, and one bedstead with black-painted posts. Arzáns (*Historia*, vol. 2, p. 160) declares that in 1654 Potosí had thirty-six gambling houses, "a great good for some people, and a great evil for others." The observation on prostitutes is from the 1603 Description of Potosí, whose author notes that in addition to these hundred and twenty women, a "great number" of Indian women followed the same trade (BAE, vol. 183, p. 379).

73. Description of Potosí (1603) (AGI Charcas 134), f. 2v.

74. See Richard M. Morse, "The urban development of colonial Spanish America," *The Cambridge History of Latin America*, vol. 2, (Cambridge: Cambridge University Press, 1984), ch. 3, pp. 68–72.

75. For this trade in ores in the *ghatu*, see Josep M. Barnadas, "Una polémica colonial: Potosí, 1579—1584," *Jahrbuch für Geschichte von Staat, Wirtschaft und Gesellschaft Lateinamerikas*, 10(1973), pp. 16—70.

76. Arzáns, *Historia*, vol. 1, p. cxxxii.

77. Bakewell, *Miners*, p. 111, and ch. 4, passim.

78. Nearly everything remains to be known about the *rancherías* of Potosí—development, population, ownership of property, administrative organization, occupations of the people, acculturative process, and so on. Archival sources are few and poor, because the dominant groups in the town, whose activities gave rise to the creation of most of the documents recording its history, had little interest in the native suburbs. Arzáns provides a starting point (see index entries under "Potosí, Villa. Rancherías . . . , in *Historia*, vol. 3, p. 548); and also useful is Luis Capoche, *Relación general de la Villa Imperial de Potosí* (ed. Lewis Hanke, BAE, vol. 122, Madrid 1959), pp. 140—41, and elsewhere.

79. See the codicil, dated Potosí, June 30, 1697, to López's will of August 9, 1694 (PCM EN 145, ff. 273—76).

80. Arzáns, *Historia*, vol. 2, p. 395.

81. PCM EN 114, f. 1,423.

82. *Auto* of don Juan de Carvajal y Sande, Potosí, February 8, 1634 (a copy remitted to Spain with Chinchón's "Gobierno No. 26" to the king, Lima, April 27, 1634. See AGI Lima 45, vol. "No. 1").

83. *Licenciado* Blas Robles de Salcedo to the Viceroy Marquis of Mancera, Potosí, April 28, 1644 (AGI Lima 52, vol. "No. 2").

84. The Viceroy Marquis of Mancera to the king, "Gobierno secular No. 15," para. 12, Lima, May 30, 1645 (AGI Lima 52, vol. "No. 3").

85. Ibid., paras. 11 and 13.

86. Ibid., attached memorandum of the *fiscal* of the Consejo de Indias, Madrid, May 2, 1646.

87. "Ministros tan grandes . . . padres de la patria y luz del mundo y sal de la tierra, que deben a los vasallos de su magestad evitar sus daños y menoscabos y darles grande ejemplo de vivir." See "Auto sobre juegos, coimas y garitos" by Nestares Marín, La Plata, November 24, 1648 (AGI Charcas 22). For Nestares's earlier career, see Jaime Contreras, *El Santo Oficio de la Inquisición de Galicia (poder, sociedad y cultura)* (Madrid 1982), pp. 204—5, 221.

88. His appointment was by *real cédula* of Madrid, August 20, 1647 (see reference in his letter "3" to the king, from Potosí, March 30, 1649, in AGI Charcas 21). For the story of earlier inspections of La Plata by

don Juan de Carvajal y Sande, and don Juan de Palacios, respectively in the mid thirties and early forties, see Schäfer, *Consejo*, vol. 2, pp. 146–49.

89. Nestares's death was reported to the king by the Viceroy Count of Alba de Aliste in his letter "No. 17" from Lima, May 28, 1660 (AGI Charcas 22). For his judgment on Potosí, see his letter to the king of March 30, 1649 (AGI Charcas 21).

90. Arzáns, *Historia*, vol. 2, p. 123.

91. *Auto* by Nestares Marín, Potosí, December 24, 1648 (AGI Charcas 22).

92. *Cabildo* of Potosí to the king, Potosí, December 1, 1649 (AGI Charcas 32, ms. 145n).

93. Arzáns, *Historia*, vol. 2, pp. 131–32.

94. The *real cédula* expressing the king's gratitude is mentioned by the Viceroy Marquis of Mancera in his "Gobierno No. 12" to the king from Lima, June 25, 1646 (AGI Lima 53, vol. "No. 4").

95. Arzáns, *Historia*, vol. 2, pp. 132–33, n. 4 by Gunnar Mendoza.

96. The Viceroy Count of Salvatierra to the king, "No. 20, gobierno secular," paras. 1–3, Lima, April 2, 1650 (AGI Lima 54, vol. "No. 3").

97. Nestares's *instrucción* from the Council, he said, "miraba a la duración de estos hombres." Ibid., para 2.

98. Para. 5 of Salvatierra's "No. 20, Gobierno secular" to the king of April 2, 1650 (see n. 96 above). The truth about this assassination attempt must remain uncertain until further manuscript evidence is found. Arzáns makes a dramatic story of it (*Historia*, vol. 2, pp. 128–29), but the archival evidence available to date is skimpy and ambiguous. No statement by Nestares himself accusing Gómez of the poisoning attempt has yet appeared, but rather only references by the viceroy (the Count of Salvatierra) to letters sent to him by Nestares, on December 31, 1649, and February 1, 1650, containing such statements. On the other hand, when Nestares wrote to the high court of La Plata on January 2, 1650, describing Gómez's offences, he mentioned nothing of the attempted poisoning, but limited himself to the general statement that Gómez, "not being able to withstand the embarrassment [*encogimiento*] in which he found himself . . . has broken out into such folly [*desatino*], undertaken with such blindness and hurry, that he finds himself without any protection or defence, or any excuse . . ." (BAN Minas t. 142, item 2[Minas catalog No. 813].

99. *Exclamación* of doña Andrea de Herrera y Bonilla, Gómez's wife, Potosí, September 16, 1649 (PCM EN 114, f. 1,405–05v.). She said of him that "por malos sucesos que ha tenido, está al presente imposi-

bilitado y falto de hacienda." The possibility cannot be discounted, of course, that her refusal to take any legal responsibility for what her husband owed was a device to safeguard family possessions.

100. Para. 5 of Salvatierra's "No. 20, Gobierno secular" to the king of April 2, 1650 (see n. 96 above).

101. Don Francisco Sarmiento de Mendoza, *corregidor* of Potosí, to the Viceroy Count of Salvatierra, Potosí, July 31, 1652 (part of an untitled *expediente* marked on the reverse "A su Magestad, No. 1, duplicado, va por Buenos Aires. Sobre la superintendencia de la Casa de Moneda que se dio a don Francisco Sarmiento de Mendoza, oidor de Lima y corregidor de Potosí . . . ," in AGI Charcas 22). The *mercaderes de plata* reported here to be in business in July 1652 were Captain Juan de Orbea, Captain Sebastián Camacho, and don Diego Calvo de Encalada.

102. Smaller coins of the same dates had their values reduced proportionately. For this order, publicly proclaimed in Potosí on May 4, 1652, see Arzáns, *Historia*, vol. 2, p. 125, n. 3, by Gunnar Mendoza. The text of the proclamation is in "Años 1652–1678. Expediente iniciado por don Francisco de Nestares Marín, presidente de la audiencia de La Plata y visitador de la Casa de Moneda de Potosí. . . . sobre el ajustamiento de la que allí se labra" (No. 657 of Mendoza's "Documentos de minas" in BAN). The fact that Nestares had allowed the striking of slightly sub-standard coin, even after his arrival, in order to keep silver traders in business, is reflected in the 6.25 percent devaluation imposed on coins made after 1649.

103. Don Francisco Sarmiento de Gamboa to the king, Potosí, July 31, 1653 (AGI Charcas 22).

104. Don Francisco de Nestares Marín to the king, Potosí, June 30, 1653 (AGI Charcas 22).

CHAPTER 2

1. Silver extracted from its ore by amalgamation emerged from the process almost pure. A mark of it was valued by Spanish law at 2,380 maravedís. The legal fineness required for coinage was 2,210 maravedís per mark. To achieve this quality, 7.5 marks of copper were added to every 100 marks of silver being prepared for minting. See "Cuenta y razón de la ganancia que tiene el que mete a hacer moneda en la casa de la moneda de Potosí," n.p., n.d., though probably August 1612 (AGI Charcas 17). This manuscript asserts that the *mercader de plata* at

that time made a maximum profit of 16 pesos 1 real in coining a bar weighing 100 marks of already taxed silver. Such a bar would cost on the open market in Potosí 826 pesos 3 reales. Assayer's and smelter's fees, wages for Indian workmen, and the cost of copper used to bring down the fineness to the legal level, added 7 pesos 4 reales to the cost of the silver itself, for a total expense of 833 pesos 7 reales. From his silver (with copper added) the merchant struck 850 pesos. The rate of profit was therefore a trifle under 2 percent. For legislation on mints and coinage, see *Recopilación* (1680), book 4, title 23.

2. The statement, although written in the third person, is by López himself, and forms part of an account of his career contained in a manuscript marked (on the reverse) "El capitán Antonio López de Quiroga, vecino de la Villa Imperial de Potosí [to] excelentísimo señor" (AGI Charcas 128). N.d., but internal evidence suggests 1670–72. López gives the year of Orbea's death in this document as 1654. But other evidence, including an earlier reference by López himself, is for 1655. In "Servicios hechos a su magestad por los capitanes Antonio López de Quiroga y Juan de Orbea, mercaderes de plata de la Casa de la Moneda de esta Villa. . . ." (in AGI Charcas 128), of which the final date is Potosí, April 29, 1668, López, writing on May 21, 1659, states that Orbea died on June 4, 1655.

3. See ch. 1, n. 101.

4. Contreras, *Santo Oficio*, pp. 205, 221.

5. A letter marked "7" in red on the first page, and on the reverse "Potosí. A su magestad. 1674. Antonio López de Quiroga. 31 de diciembre" (AGI Charcas 128).

6. Ibid. López states that Nestares reduced the fees paid by the traders to the treasurer and other officials of the Mint, and also the amount deducted for the crown under a tax called the "derecho de Cobos." This was actually a levy of 1.5 percent, applicable to all silver produced, that was charged at the Treasury office in Potosí on bar silver presented there for taxation. Exactly what Nestares did with the "derecho de Cobos" is hard to follow. Initially, on August 28, 1652, he removed it on bars of silver that were due to be coined in the Potosí Mint (PCM CR 316, ff. 148–48v., "Auto del Señor Presidente en razón de la quita [d]el uno y medio por ciento de Cobos, que llama, y la imposición sobre las barras.") But that, apparently, was insufficient incentive to the silver brokers. So further amounts were cut in 1652 and early 1653 from the tax (presumably from what was being levied on bars *not* due for minting). By March 1653 the imposition had been reduced by a total of 40,000 duc-

ats (c. 55,000 pesos) a year (PCM CR 316, Libro de Provisiones 1649–53, ff. 170–70v., 174v.–75). The absence of references to "Cobos" in the Potosí treasury accounts between 1652 and 1684, however, suggests that the tax was not collected at any level in that period. That impression is given strongly also in "Borrador de un informe por don Bartolomé González de Poveda, presidente de esta Real Audiencia, para el Duque de la Palata, virrey del Perú, sobre el origen del derecho llamado de Cobos . . .," of La Plata, March 2, 1684 (BAN Minas t. 135, item 5 [Minas catalog No. 1,058]), f. 16.

7. For López's presentation of bar silver for taxation at the Treasury in 1653, see the statement by Antonio Cupín de Esquivel, accountant (*contador*) of the Treasury, of Potosí, August 22, 1659, in "1672 [for 1662–1689. Francisco Guerra Zavala, receptor de alcabalas en Potosí, sobre lo que deben por ese derecho en razón de sus tratos y contratos el capitán Pablo de Espinosa Ludueña, el maestre de campo Antonio López de Quiroga, y el capitán Antonio de Cea. . . ." (BAN Minas t. 136, item 2 [Minas catalog No. 1,190]), f. 94. The evidence for López's starting coinage operations in the Mint in April 1655 is from a *memorial* that he and Antonio de Cea sent to the Viceroy Count of Alba de Aliste in 1657 (no exact date) (ibid., f. 10v.). The first independent evidence of his minting activities dates from the beginning of the next year. On January 11 and 19, 1656, López received 85 bars of silver at the Treasury, worth 54,957 pesos 5 reales 5 granos (*plata ensayada* of 425 maravedís per peso) for conversion into coin at the Mint. He paid for these bars in coin on March 13, 1656 (PCM CR 361, Libro Real Común for 1656, "Cargo de la plata procedida de las barras que se vendieron para hacer reales," f. 70v.).

8. *Provisión* by the Viceroy Count of Alba de Aliste, "Sobre el peso de las barras que se labran moneda," Lima, December 31, 1657 (PCM CR 339, Libro de provisiones 1653–58, ff. 184v.–85v.).

9. See three separate *recibos de plata* issued to Antonio López by Pedro de Trujillo, a resident of Potosí departing for Lima, on October 19 and 22, 1654. The shipments of silver were to be delivered to Captain Domingo Montero Salarind[is?], don Pedro de Loaysa y Quiñones, and the *Tesorero* Luis López de Chauri (respectively, PCM EN 116, f. 935–35v., 1,008–8v., and 1,011–11v.).

10. See the remarks in ch. 1 on the close ties between Quirogas and Losadas. For the debt collection, see the *cesión* of Potosí, September 28, 1654, issued by Liano de la Vega to Antonio López (PCM EN 116, f. 898–98v.), and the *poder* of the same date granted by López to Losada y

Novoa and Graviel de Moscoso (*vecino* of Cuzco) to collect the sum owed from Vargas (ibid., f. 899–99v.).

11. *Carta de pago* issued by Somoza Losada y Quiroga to Antonio López, at the asiento del Espíritu Santo, "minas de la provincia de Carangas," December 26, 1655 (BAN Minas t. 144 [Minas catalog No. 852], f. 1,334–34v.). On another occasion, López collected income from the *encomienda* of the province of Atacama for remittance to Lima. He received 5,053 pesos from the *encomendero* of Atacama, the *maestre de campo* don Baltasar Salgado de Araujo and delivered them to a priest named Tomé Luis de Comba for carriage to Lima. See Comba's *carta de pago* to López, of Potosí, June 30, 1654 (PCM EN 116?, folio not recorded).

12. Fray Antonio del Puerto, O.F.M., "Certificación de cómo el maestre de campo Antonio López de Quiroga es síndico de la religión de Nuestro Santo Padre San Francisco," Potosí, January 26, 1689 (AGI Charcas 128).

13. Juan Sueiro Leytón to Antonio López de Quiroga, for the convento de San Francisco, "Obligación de entregar madera," Potosí, June 15, 1654 (PCM EN 116, ff. 493–94).

14. He named a member of the Franciscan community in Potosí, Fray Juan Osorio, a man "very suitable for the said purpose," to replace him. See his *poder general* to Osorio of Potosí, November 15, 1670 (PCM EN 123, ff. 596–97v.). Despite this, he can still be seen issuing receipts for sums received by the monastery in 1673 and 1677 (PCM EN 125 and 127, passim).

15. Statement of Fray Buenaventura de Madrid, *guardián* of the Franciscan house in La Plata, La Plata, January 7, 1690, in "1689–1690. Información de los méritos y servicios del maestre de campo Antonio López de Quiroga en las labores de minas en Potosí, Porco, Chayanta, Lipes y otros parajes de estos reinos" (BAN Minas t. 19, item 2 [Minas catalog No. 1,100]), f. 14.

16. Fray Antonio del Puerto, O.F.M., "Certificación de cómo el maestre de campo Antonio López de Quiroga es síndico de la religión de Nuestro Santo Padre San Francisco," Potosí, January 26, 1689 (AGI Charcas 128).

17. ". . . [C]on que perdí más de 870,000 pesos incobrables hasta hoy" [1674]. See letter "7" of Potosí, December 31, 1674 in AGI Charcas 128 (as in n. 5 above). For the 150,000 peso loss under Orbea's administration, see the initial *petición*, dated Potosí, May 21, 1659, in "No. 4 A. Servicios hechos a su magestad por los capitanes Antonio López de

Quiroga y Juan de Orbea, mercaderes de plata de la Casa de la Moneda de esta Villa, de barras que han presentado a quintar ... desde 20 de diciembre del āno de 1650 hasta 29 de abril del de 1668. ..." (AGI Charcas 128).

18. Ibid. For the lack of rain in 1651, see the testimony of Pedro de Ballesteros, of Potosí, April 30, 1652, in an *Información* about the restoration of the reservoirs of Potosí undertaken by don Francisco de Nestares Marín (AGI Charcas 21); and also that of don Andŕes de Sandoval, royal treasurer in Potosí, to the effect that the drought, lasting five months, was the worst in Potosí for thirty years (ibid.). The dryness seems to have continued until the end of 1652, although by the end of February 1653 so much rain had fallen that it was reported that refiners had difficulty drying ore enough to mill it. See Treasury officials of Potosí to the Viceroy Count of Salvatierra, December 30, 1652 and February 28, 1653 (PCM CR 503, ff. 30v., 35).

19. See n. 17 above.

20. The title appears in a commercial "Consiento y obligación" he drew up in Potosí on April 22, 1659 (PCM EN 118, f. 464).

21. "Declaración. Juan Bautista de Bastarrica en favor del capitán Antonio López de Quiroga," Potosí, May 17, 1661 (PCM EN 199A, f. 316–16v.).

22. He refers to the sale (of March 23, 1658) in a later document: "Poder. El maestre de campo Antonio López de Quiroga a don Joseph de Asáldegui," Potosí, August 29, 1696 (PCM EN 144, f. 403). The seller was one Captain Alonso de Fonseca Falcón.

23. Statement, Potosí, August 22, 1659, of Antonio Cupín de Esquivel, accountant of the Treasury, in "Francisco Guerra Zavala ...," f. 93 see n. 7 above).

24. Ibid.—the royalty paid on what López and Orbea presented to the Treasury came to 1,849,269 pesos ensayados 5 reales (pesos of 450 maravedís). The total royalty paid on all silver registered at the Treasury for 1651–59, inclusive, was 4,537,769 pesos ensayados (Bakewell, "Registered silver production ...," p. 95). López's amount is 40.75 percent of the total. The percentage would rise slightly with any deduction made from the total for the final four months of 1659, since there is no information on López's deliveries of silver to the Treasury for September–December 1659.

25. PCM CR 321, f. 201. A royal mine was by law set aside, adjacent to the first claim registered on any new vein, by the discoverer of the vein. Such mines were always leased out by the Treasury in seventeenth

century Potosí. The lease to López does not say whether the mine had previously been worked.

26. Ibid., f. 102, lease of April 25, 1657 of the royal mine in the vein of San Antonio de Padua, registered by Luis Lobo.

27. Letter "7" of Antonio López, Potosí, December 31, 1674 (AGI Charcas 128).

28. See the rough plan of the Cerro of Potosí, showing names of principal veins, accompanying the letter of the *Licenciado* Juan de Torres to the king, Potosí, March 1, 1620 (AGI Charcas 52). I am indebted to Srta. Clara López Beltrán for copies of the letter and plan.

29. "71. 1675–1676. El maestre de campo Antonio López de Quiroga con Juan de Figueroa y consortes, sobre el derecho al socavón llamado Nombre de Jesús, o Mala Moneda, paraje de las Amoladeras, cerro de Potosí" (BAN Minas t. 16, item 1[Minas catalog No. 1,003]), ff. 226v.–27.

30. "Consiento y obligación" between don Diego Fernández, of Jujuy, and Captains Francisco de Eyzaguirre and Antonio López de Quiroga, Potosí, April 22, 1659 (PCM EN 118, f. 464–64v.).

31. He bought the refinery from doña Catalina de Sedeño de Contreras for 6,991 pesos. The date of the sale is not recorded, but seems to have been before 1660. In 1652, the refinery had been appraised at 13,000 pesos. See BAN Minas t. 12, item 6 [Minas catalog No. 906], ff. 101v.–2v., and the continuation of litigation in ibid., item 7 [catalog No. 919 ff. 71, 120.

32. BAN Minas t. 12, item 6, ff. 25v.–26v., inventory taken of Uceda's refinery, Potosí, December 1, 1658.

33. Ibid., ff. 11v.–13, 14–16v., 23, 35–52; and BAN Minas t. 12, item 7, ff. 118–23.

34. See n. 31 above.

35. BAN Minas t. 16, item 1, f. 69.

36. For López's mercury purchases in the 1660s, see PCM CR 384, f. 175v. In the decade 1656–65, an average of 120.1 marks, or 60.05 pounds, of silver was produced with the loss, or "consumption," of 100 pounds of mercury (see Bakewell, "Registered silver production . . . ," Table 2, p. 98). For total registered silver production, ibid., p. 95.

37. In 1665 the number of *azogueros* was counted at fifty-nine. See f. 22 of an unsigned "Información" on the number of *mita* Indians distributed to refiners, dated Lima, July 25, 1665, in AAGN Sala 13, Cuerpo 23. 10–2, cuaderno 7.

38. BAN Minas t. 16, item 1, ff. 69–74v.

39. Ibid., ff. 76–84, 211–12. For the regulations on how many work-

ings might be held by one person, see ordinances 12, 13, and 14 in the first title ("De los descubridores, registros y estacas") of *Ordenanzas del Virrey don Francisco de Toledo acerca de los descubridores, registros y estacas de las minas; de las demasiás, medidas y amojonamientos, cuadras, labores y reparos, entradas de unas minas en otras, despoblados, socabones, alcalde de minas, determinación de pleitos; desmontes, trabajo y pago de los indios*, issued at La Plata, February 7, 1574. For this set of ordinances, see Roberto Levillier, *Gobernantes del Perú. Cartas y papeles,. siglo XVI. Vol. 8. Ordenanzas del Virrey Toledo* (Madrid 1925), pp. 143–240. The discoverer of a vein might register two claims in it. No one else might have more than one. Toledo, indeed, imposes in ordinance 14 an upper limit of five mines per person, each of which should be in a different vein. This regulation was certainly no longer in force in Antonio López's time.

40. PCM CR 384, ff. 336v.–39.

41. BAN Minas t. 16, item 1, f. 57–57v., donation, Potosí, January 13, 1662.

42. Ibid., f. 60.

43. For the dimensions of the adits, "Declaración de los alcaldes veedores," Potosí, March 5, 1676 (ibid., f. 187). López, late in 1675, maintained that his workings at Amoladeras had no fewer than twelve access points on the surface, though some of them were not being used (ibid., f. 64–64v.). The president's observations are in "La Plata. Duplicado a su magestad 1682. Don Bartolomé González Poveda que fue presidente, 20 de agosto, informa los méritos del maestre de campo Antonio López de Quiroga" (AGI Charcas 128).

44. Testimony of Captain Joseph Molero, *mercader*, Potosí, December 1, 1689, in an "Información" by Antonio López de Quiroga, Potosí, January 21, 1690 (AGI Charcas 128), f. 26v.

45. *Historia*, vol. 2, p. 395. Arzáns expresses the yield as 800 marks per *cajón* of 50 quintales of ore.

46. BAN Minas t. 16, item 1, f. 108v., *petición* of don Juan de Figueroa, Potosí, 1675 (no exact date).

47. "Tan poderoso y valido"—in contrast with the poverty of the plaintiff, don Juan de Figueroa, and his wife, doña María de Mesa Figueroa (daughter of the doña Paula de Figueroa who had given Antonio López the adit "Mala Moneda" and attached mines in 1662). See the *interrogatorio* of La Plata, May 12, 1676, addressed by Figueroa to the high court of La Plata, question 10 (ibid., f. 263).

48. For mercury entering Potosí, 1661–69, see PCM CR 385, 393, 399,

402, 408, 411, 415, 419, and 423. For López's mercury purchases in those same years, see PCM CR 384, f. 175v. It is assumed here, in making López's percentage share of mercury the same as his share of silver output, that he produced the same amount of silver from a given weight of mercury as did the average refiner in Potosí. This may not have been so, but in the absence of any means of knowing if he did better or worse than the average, it is the best assumption to take. The average output of silver per 100 pounds of mercury "consumed" rose in Potosí from 60 pounds in the decade from 1657 to 1666 to 715 pounds in the decade from 1667 to 1676. (Bakewell, "Registered silver production . . . ," Table 2, p. 98.) To take account of this increase, the calculation of López's silver output of 200,000 pounds for 1661 to 1669 given in this paragraph is based on a ratio of 65 pounds of silver produced per 100 pounds of mercury consumed.

49. "Información" of July 25, 1665 (see n. 37 above).
50. PCM CR 384, ff. 173v.–74.
51. Ibid., f. 175v.
52. Ibid., ff. 188v.–89.
53. PCM CR 445, f. 135v.
54. Ibid., f. 139
55. *Compañía* of López and Ibáñez, Potosí, January 29, 1669 (PCM EN 121, ff. 82–84v.).
56. Testimony of Ambrosio Ruiz, Potosí, November 23, 1689, in "Información" by Antonio López, 1690, f. 11v. (as in n. 44 above).
57. *Historia*, vol. 2, p.396.

CHAPTER 3

1. Don Diego Muñoz de Cuéllar, Juan de Urdinzu y Arbeláez, Gaspar de Torres, and Antonio López to the king, Potosí, August 31, 1668 (AGI Charcas 36).
2. *Historia*, vol. 2, pp. 395–96. "Dióle una palmadita en las espaldas diciéndole: 'Muy rico estás, Antuco', y luego quiso doblar la correspondencia con mayor garbo pues le dijo: 'Sabed que la condesa vuestra paisana está encinta y quiero haceros mi compadre: mirad qué disposición tenéis para esto'."
3. It was born on or about July 10, 1668. See Guillermo Lohmann Villena, *El Conde de Lemos, virrey del Perú* (Madrid 1946), pp. 201–02.
4. *Historia*, vol. 2, p. 396.

5. Lohmann Villena, *El Conde de Lemos*, chs. 11–14. See also Meredith D. Dodge, *Silver mining and social conflict in seventeenth-century Peru: Laicacota, 1665–1667* (Ph.D. dissertation, University of New Mexico, 1984).

6. Lohmann Villena, *El Conde de Lemos*, p. 211.

7. In another document of fifteen years later, López did record that he had gone to Laicacota as a representative of the *azogueros*, accompanied by don Diego Muñoz de Cuéllar in the same capacity. See his "Información" of January 21, 1690 (AGI Charcas 128), f. 2, question 4.

8. Letter "7" by Antonio López to the king, Potosí, December 31, 1674 (AGI Charcas 128). The sack of Porto Belo that López alludes to was the work of the English pirate, Henry Morgan, on July 12, 1668. News of it reached Lima on August 31. See Lohmann Villena, *El Conde de Lemos*, p. 342.

9. For Lemos and the *mita* of Potosí, see Cole, *The Potosí mita*, pp. 97–103.

10. In Antonio de Valenzuela to "muy poderoso señor," Potosí, April 1, 1690 (AGI Charcas 128), testimony of the *maestre de campo* don Agustín Ortega y Retuerta, La Plata, January 12, 1690.

11. PCM CR 418, f. 142v. López was required to pay the nominal sum of 500 pesos for title to the adit and mines.

12. For the *bandidos*, letter "7" by López to the king, Potosí, December 31, 1674 (AGI Charcas 128). For the *corregidor*'s report, Antonio López to the king, Potosí, December 4, 1679 (a holograph letter marked "12") (AGI Charcas 128), f. 2v.; for his expenditure, the "Informatión" of January 21, 1690 (AGI Charcas 128), f. 2, question 4.

13. Testimony of Dr. don Pedro Vázquez de Velasco, treasurer of the Cathedral of La Plata, January 3, 1690, in "1689–1690. Información de los méritos y servicios del maestre de campo Antonio López de Quiroga . . . ." (Ban Minas t. 19, item 2 [Minas catalog No. 1,100]), f. 6.

14. Letter "7" of Antonio López, Potosí, December 31, 1674 (AGI Charcas 128).

15. Don Bartolomé González Poveda, president of the Audiencia of La Plata, to the king, Potosí, August 20, 1682: "Informe de los méritos y servicios del maestre de campo [Antonio López de Quiroga], azoguero de la Villa Imperial de Potosí" (AGI Charcas 128), f. 2v.

16. Antonio de Valenzuela to "muy poderoso señor," Potosí, April 1, 1690 (AGI Charcas 128). Also, "Información" of Antonio López, Potosí, January 21, 1690 (AGI Charcas 128), f. 20v., question 6, testimony of

Captain Blas Míguez (a former *mayordomo mayor* of Antonio López's mines and refineries), Potosí, November 24, 1689.

17. For blasting at Huancavelica, see Eugenio Maffei and Ramón Rúa Figueroa, *Apuntes para una biblioteca española de libros, folletos y artículos, impresos y manuscritos, relativos al conocimiento y explotación de las riquezas minerales y a las ciencias auxiliares* (2 vols., Madrid 1871, repr. VI Congreso Internacional de la Minería, León 1970), vol. 1, p. 485 (item 1,680). The reference is to a manuscript treatise of 1642 entitled *Memorias antiguas y nuevas del Perú* by the *Licenciado* don Fernando Montesinos, who observes that the adit of Huancavelica was begun in 1609, and was carried forward until 1635 by hand work. At that point, blasting began, by which it was advanced more in four years than in the first twenty-six. Maffei and Rúa Figueroa (vol. 1, p. 485, n. 1) note that blasting was not introduced into mining in Spain until 1703. The first successful application of blasting to mining in Europe seems to have been made in 1627, at Schemnitz in Hungary. "Perhaps the most telling reference would be in the Proceedings of the Schemnitz Mine Tribunal . . . for February 8, 1627, where Caspar Weindl is considered as taking an unfair advantage over his fellow miners by his adaptation of gunpowder to mining. Apparently he used the gunpowder in bore holes. . . ."—personal communication to the author from Mr. Brian Earl, Sennen, Cornwall, U.K., of February 26, 1986.

18. William Hooson, *The miner's dictionary, explaining not only the terms used by miners, but also containing the theory and practice of that most useful art of mining, more especially of lead mines . . .* (Wrexham 1747, repr. by The Institution of Mining and Metallurgy, London 1979), entry for "blasting."

19. Ocurí is on the present Sucre-Oruro road, and Aullagas a little to the northwest of it, close to Colquechaca. Titiri appears on no modern map known to the author, but documentary references strongly suggest it was near the other two sites.

20. *Reales oficiales* of Potosí to the Viceroy Count of Alba de Aliste, Potosí, June 30, 1656 (PCM CR 503, f. 64v.). For the assertion of discovery by López and Cea, "1662–1689. Francisco Guerra Zavala . . .," (BAN Minas t. 136, item 2 [Minas catalog No. 1,190]), f. 10v.

21. Don Bartolomé González Poveda to the king, La Plata, June 30, 1680 (AGI Charcas 24).

22. For the Aullagas adit in the early seventies, "No 2." [on reverse] "El capitán Antonio López de Quiroga de la Villa Imperial de Potosí [to] excelentísimo señor," no date (AGI Charcas 128). The later report is a

holograph letter of Antonio López to the king, Potosí, September 28, 1679 (AGI Charcas 128).

23. "Transacción, venta, cesión y traspaso" by Francisco de Viveros to Antonio López de Quiroga, Potosí, July 23, 1675 (PCM EN 127, ff. 131–37v.). For López's mining property at Ocurí, "1679. El maestre de campo Antonio López de Quiroga, dueño de minas e ingenio en el asiento de Ocurí, provincia de Chayanta, sobre las cuentas de la administración de dichas haciendas que tuvo a su cargo Antonio Largáñez" (BAN Minas t. 70, item 3[Minas catalog No. 1,029]), f. 1. No documentary references to Ocurí are to be found dating from before 1649, whi⁻h suggests that it had not long been active at that time. See, for example, the "Concierto y obligación" of Potosí, May 8, 1649, between Juan de Segura, a carpenter in Potosí, and don Diego Guillén del Castillo, with Andrés Ortuño de la Rea, for the building of a water-driven silver refining mill half a league from Ocurí (PCM EN 112, ff. 718–19v.); or the "Deudo" of Salvador de Escalona, a resident in the *asiento* of Ocurí, and owner of a small ore mill (*trapiche*) there, of June 15, 1649, acknowledging a loan of 1,736 pesos from Gervasio Navarro, *alcalde provincial* of the Santa Hermandad in Potosí (PCM EN 114, ff. 895–96).

24. BAN Minas t. 70, item 3, f. 1–1v.

25. López to Largáñez, Potosí, August 13, 1674 (ibid., ff. 13v.–14).

26. *Memorial* by Largáñez of his receipts and expenditures (ibid., ff. 6v.–8v.). It would seem from these figures that López was losing heavily at Ocurí, but it is unclear whether this was the gross silver production, or, as Antonio López maintained, net output after various costs had been paid. This in fact became a point of dispute between López and Largáñez. For the standard of equipment of López's refineries, see question 1 of the testimony of don Diego de Figueroa y Paz, Potosí, November 22, 1689, f. 6 in the "Información" on Antonio López of Potosí, January 21, 1690 (AGI Charcas 128).

27. López to Largáñez, Potosí, July 7, 1674 (ibid., f. 14v.).

28. López to Largáñez, Potosí, August 18, 1676 (ibid., f. 15).

29. López to Largáñez, August 13, 1674 (ibid., f. 14).

30. Testimony of don Diego de Figueroa y Paz, Potosí, November 22, 1689 ("Información" on López of January 1690, as in n. 26 above).

31. *Interrogatorio*, question 4, and testimony of don Diego de Figueroa y Paz and Captain Ambrosio Ruiz de Villodas, ff. 1v., 7, and 15v. respectively, ibid.

32. One report puts discoveries of ores there in 1647. See the testimony of Captain Blas Durán de Montalbán, Ocurí, January 8, 1675,

in "1674–75. El maestre de campo Antonio López de Quiroga con los hermanos Francisco, Diego, y José de Núñez, sobre el derecho a la mina nombrada la Venturosa, asiento del Espíritu Santo de Titiri, provincia de Chayanta" (BAN Minas t. 60, item 1 [Minas catalog No. 1, 000]), f. 162.

33. Ibid., especially ff. 19v., 34v., 39v.–42, 106v., 115–16, 118, 128, 142v., 164v., 252. For the refinery in 1690, Antonio de Valenzuela to "muy poderoso señor," Potosí, April 1, 1690.

34. Fray Buenaventura de Madrid, *guardián* of the Franciscon monastery in La Plata, testimony of La Plata, January 7, 1690, in "1689–1690. Información de los méritos y servicios del maestre de campo Antonio López de Quiroga . . ." (BAN Minas t. 19, item 2 [Minas catalog No. 1,100]), f. 12v.

35. Geographically speaking, Oruro can be said to have lain in the district of Potosí, and it was certainly, like Potosí, in the jurisdiction of the high court of La Platâ. On the other hand, since it possessed its own Treasury office from 1607 onward, in fiscal administration Oruro was a separate entity from Potosí. And this was sufficient to set it apart as a small mining district in its own right.

36. For Berenguela, Alvaro Alonso Barba, *Arte de los metales, en que se enseña el verdadero beneficio de los de oro y plata por azogue* ([Madrid 1630], ed. used, Potosí 1967, pp. 9–10; and for an early reference to Tomahavi, a *Donación* of part of the first mine registered there (named Nuestra Señora de la Asunción) by Pedro Pérez Hidalgo to Alonso de los Reyes, Potosí, September 29, 1614 (PCM EN 47B, f. 2,010–10v.).

37. The adit at Berenguela was under way by January 1675. See "1675. El maestre de campo Antonio López de Quiroga, con Bartolomé de Villalobos, sobre el derecho a la veta nombrada San Francisco de Asís y un socavón que se está dando hacia ella, en el asiento de Berenguela, provincia de Pacajes" (BAN Minas t. 97, item 5 [Minas catalog No. 999]). For Tomahavi, letter "12" of Antonio López to the king, Potosí, December 4, 1679 (AGI Charcas 128), f. 2v.—where López states that three years' work have gone into the adit (though it is not clear when this effort began).

38. Besides the adits, mines, and refineries in Chayanta that have already been mentioned, López was evidently planning in 1671 to build a further refinery close to a mining settlement called Orcopata, in the same district. The mill (water-driven) was to stand on land bordering the jurisdiction of the Indian town of Chayanta itself, and López agreed to pay 100 pesos to don Juan Payna, governor of the town of Aymaya in

the province of Chayanta, in return for permanent use of the site for a refinery. Absence of later references to this mill suggests, however, that in the event it was not built. See the petition of Ambrosio Ruiz de Villodas, on behalf of Antonio López, Potosí, April 25, 1671 (PCM EN 124, ff. 186–89v.).

39. "Obra singular en todo el Reino"—from a memorandum prepared by Pedro de Vergara y Borda and Juan Esteban de Legama in response to a royal order to the Count of Adanero: "Sobre lo que suplica en el memorial incluso el maestre de campo Antonio López de Quiroga, minero y azoguero de la Villa de Potosí, me informaréis lo que se os ofreciere y pareciere," Madrid, August 19, 1697 (AGI Charcas 128).

40. "Carta que escribió Juan Lozano Machuca, factor de la Nueva Toledo, al virrey don Martín Enríquez, sobre las minas y riqueza de los Lipes," Potosí, November 26, 1581 (AGI Charcas 35, item 51).

41. Ibid. The writer does not say what the Aymaras' crops were, but they are likely to have been potatoes and *quinoa*, an Andean grain, rich in protein, related to pigweed.

42. Don Juan de Lizarazu to the king, Potosí, March 8, 1635 (AGI Charcas 20).

43. Fifteen "interesados en minas" to "vuestra alteza," Asiento de San Antonio, November 30, 1647 (BAN Minas t. 56, item 4 [Minas catalog No. 771]).

44. "Fletamento" of December 24, 1648, undertaken by Pedro Fernández Cerdán for Joseph Maldonado. The amount to be shipped was 400 *costales* (sacksful) of 6 arrobas, 5 pounds the sack. Fernández was to receive the cargo by January 3 or 4, 1649, and deliver it in San Antonio between February 1 and 12, at a rate of 7 pesos per *costal*.

45. Treasury officers to the Viceroy Count of Alba de Aliste, Potosí, June 30, 1657 (*primera carta*) (PCM CR 503, ff. 73v.–74). Of the royalty on silver production of a fifth collected in Potosí between January 1, 1655 and June 1, 1657 for remittance to Spain (a total of 1,245,882 *pesos ensayados* of 450 *maravedís*), 49.82% had come from the district (*contorno*). Of the 1,211,096 *pesos ensayados* in royalty collected for despatch in the previous three *armadas*, or fleets to Panama, 51.02% came from the district. When precisely those fleets sailed, the Treasury officers do not say. In all probability, however, they were referring to remittances sent off from Potosí in 1652, 53, and 54.

46. Bakewell, "Registered silver production," Table 3, p. 99.

47. See the reference to this letter (of November 12, 1676) in a *real*

cédula from the king to Castellar, Madrid, January 14, 1678 (AGI Charcas 128).

48. Letter "7" of Antonio López to the king, Potosí, December 31, 1674 (AGI Charcas 128).

49. For the purchase price of his office, see "1684. Los oficiales reales de Potosí sobre los 6,000 pesos de censo que don Alvaro de Espinosa Patiño . . . impuso en dicho oficio a favor de la real hacienda" (BAN Minas t. 136 [Minas catalog No. 1,059]), f. 1. Espinosa's parents were Captain Pablo de Espinosa Ludueña and doña Ana María Patiño de Velasco—see extracts from his will, Potosí, February 24, 1687 (BAN Minas t. 60, item 5 [Minas catalog No. 1,122], ff. 41v.–43.).

50. In 1662 Antonio López had paid 3,000 pesos into the Treasury in Potosí on behalf of Captain don Cristóbal de Quiroga, almost certainly the same man as Cristóbal de Quiroga y Osorio, since, among other things, he was associated with los Lipes at that time (PCM CR 393?, f. 121–21v.).

51. In 1678 López gave him a wide power-of-attorney in los Lipes: to administer refineries, adits, and mines; to secure *avío* and labor for all López's enterprises there; to keep accounts; and to claim new mines (PCM EN 130, ff. 190–92v.).

52. See Ordenanza 10 of the title "De los socavones" of don Francisco de Toledo's mining ordinances (La Plata, February 7, 1574), in Roberto Levillier, *Gobernantes del Perú. Cartas y papeles, siglo XVI* (14 vols., Madrid 1921–26), vol. 8, p. 213.

53. "1679–1681. El maestre de campo Antonio López de Quiroga contra don Alvaro de Espinosa Patiño, sobre la parte que a éste le toca en los gastos del socavón que ambos daban en el asiento de San Antonio del Nuevo Mundo, provincia de los Lipes" (BAN Minas t. 58, item 5 [Minas catalog No. 1,042]), f. 70 (part of a *Memorial* by Captain Alonso Ruiz, Potosí, July 3, 1678).

54. Ibid.

55. Ibid., ff. 48, 261.

56. Ibid., f. 69v.

57. Ibid.

58. Ibid., f. 70; for the dimensions and linkages, f. 248.

59. "Declaración de los veedores," San Antonio del Nuevo Mundo, November 23, 1679 (BAN Minas t. 19, item 1 [Minas catalog No. 1,088]), ff. 61v.–63v.

60. BAN Minas t. 58, item 5 (for title, see n. 53 above), f. 266; for the drainage channel, f. 222v.

61. José de Acosta, *Historia natural y moral de las Indias, en que se tratan [sic] de las cosas notables del cielo, elementos, metales, plantas, y animales dellas, y los ritos y ceremonias, leyes y gobierno de los indios (1590)*, ed. Edmundo O'Gorman (2nd. ed., Mexico City 1962), Libro Cuarto, Capítulo 8 (p. 156).

62. "Durezas incontrastables"—see BAN Minas t. 19, item 1, f. 58v., *Petición* of Alonso Ruiz, no place cited, November [?], 1679. An obvious question arising from Ruiz's animated self-congratulation about the good effects of blasting is: Where did all the powder come from? His summary accounts give no hint of a reply—except to suggest that López was not manufacturing powder himself, since entries for powder purchases are given as outgoings, with variable prices, in exactly the same way as purchases of tools and other items are recorded. Powder, however, was clearly made in Potosí and in La Plata throughout the seventeenth century for use in fireworks and guns. See, e.g., Arzáns, *Historia*, vol. 1, p. 367, and vol. 2, pp. 256–57. Furthermore, Spaniards had known from well back into the sixteenth century that deposits of saltpeter (potassium nitrate, the main ingredient of gunpowder) existed in los Lipes. See the "Carta que escribió Juan Lozano Machuca . . . ," Potosí, November 26, 1581 (n. 40 above). The other two ingredients of powder were charcoal, of which Potosí had no shortage, and sulphur, which might have been taken, as Cortés took it during the conquest of the Aztecs, from the interior of Popocatépetl, from volcanoes lining the *altiplano*. For all this, nevertheless, the adoption of blasting in mining in the Potosí district during the 1670s must have raised the prevailing demand for powder enormously, and an exploration of how and where this new raw material was made would be a welcome piece of research.

63. Ibid. The cartridges themselves were made of sheepskin, and sometimes wax was used to seal them instead of pitch. See BAN Mines t. 58, item 5, ff. 65, 67, 259.

64. This amount of powder, at the average price given by Ruiz, would have cost 11.8 pesos *de a ocho*. This figure might serve as the point of departure for calculating the relative costs of tunneling with and without blasting. At the time of writing, however, insufficient information is available on the cost of tunneling with hand labor alone for that comparison to be completed. Differences in hardness of rock may also make comparison drawing a dubious enterprise, at least until a large number of cases are gathered. It is conceivable, for example, that the rock that Ruiz cut into at San Antonio was so hard that tunneling with hand work alone would have been so slow and expensive as to have been impracti-

cable. For *barretas* and the weight of charges, see BAN Minas t. 58, item 5, f. 224.

65. Castellar to king, "13," Lima, March 15, 1678 (AGI Charcas 123).

66. Bakewell, "Registered silver production," Table 3, p. 99.

67. Ibid.

68. See the account entry for "La tropa de los indios que trajo Carlos de Quiroga y Bartolomé Gallegos para sacar metales" in 1671, in BAN Minas t. 58, item 5, f. 62v. The connection between Carlos de Quiroga and Antonio López is not established, though he was evidently at least an employee or agent. In 1688 he reappears as a resident of Potosí, and with the title of Captain, buying a silver refinery in the province of Chayanta from a prominent citizen and *mercader de plata* in Potosí, Captain don Lorenzo de Narriondo y Oquendo (PCM EN 137, ff. 347–48v., purchase agreement, Potosí, October 29, 1688).

69. The lease was from the Treasury, to which a previous tenant owed money for failure to pay his rent, and had therefore been removed. During López's lease he was to rebuild the *ingenio*, and at the end return it in running order (PCM CR 445, ff. 136, 139).

70. PCM CR 360, f. 82v.

71. BAN Minas t. 58, item 5, ff. 64–69.

72. BAN Minas t. 19, item 1, ff. 46, 54, 56–56v.: statements by Ruiz and Figueroa (*procurador de causas del número* in Potosí), various dates, 1683.

73. Decision of La Plata, January 28, 1689 (ibid., f. 117).

74. "Astucias"—see BAN Minas t. 60, item 2 [Minas catalog No. 1,105], no precise date, but January 1691. For Ruiz's *ingenio* in San Antonio, BAN Minas t. 60, item 2 [Minas catalog No. 1,105], f. 42.

75. "No aresgó [*sic*] sólo un peso en el socavón en la compañía a que nos obligamos a dar dicho socavón. . . ." López to the king, "12," Potosí, December 4, 1679 (AGI Charcas 128), f. 1v.

76. "Mas esto de gastar lo que tienen en casa hay muy pocos que lo hagan." López to the king, Potosí[?], September 28, 1679 (AGI Charcas 128).

77. López to the king, "12," Potosí, December 4, 1679 (AGI Charcas 128), f. 1v.

78. A copy of the agreement, dated Potosí, May 15, 1687, is with a report from the President of La Plata, don Diego [Cristóbal] Mesía to the king, La Plata, September 3, 1690 (AGI Charcas 28), f. 3–3v. The report is marked on the verso "Refiere el buen efecto que ha tenido la com-

posición que hizo [Mesía] en el pleito que seguía Antonio López de Quiroga con los albaceas de don Alvaro de Espinosa Patiño . . . ."

79. "Este es el mayor servicio que ahora de presente puedo hacer a vuestra Magestad, aunque son muy grandes los hechos hasta hoy, que no se contarán de otro vasallo. . . ." López to the king, Potosí[?], September 28, 1679 (AGI Charcas 128).

80. López to the king, Potosí, August 12, 1690 (AGI Charcas 128), f. 1. If the distance of 1,470 yards referred to the main underground gallery of the adit alone, then this had been almost doubled in length since 1678. The Santo Domingo vein must also, in that case, have been much farther from the limit of the working in 1678 than López then thought.

81. Don Diego [Cristóbal] Mesía to the king, La Plata, September 3, 1690 ("Refiere el buen efecto . . ."), f. 1v. (AGI Charcas 28). For output registered at the Treasury in Potosí in 1690, see Bakewell, "Registered silver production", Table 3, p. 100.

82. Testimony of *maestro* don Juan Araujo, La Plata, January 5, 1690, in Antonio de Valenzuela to the king, Potosí, April 1, 1690 (AGI Charcas 128). See also ibid., testimony of Dr. don Pedro Vázquez de Velasco, La Plata, January 3, 1690.

83. See for these problems BAN Minas t. 60, item 5 [Minas catalog No. 1,122]: "1693–1694. El maestre de campo Antonio López de Quiroga sobre que se le adjudiquen las vetas Ancha, Rica y de la Concepción, asiento de San Antonio del Nuevo Mundo, provincia de los Lipes, para desaguarlas," especially ff. 1–1v., 15–18.

84. Ibid., ff. 26–28.

85. Bakewell, "Registered silver production." Table 3, p. 100.

86. "1703. Copia de una relación de don Juan Antonio de Roiz y Balcázar, corregidor de la provincia de los Lipes, sobre el miserable estado en que la encontró por la cortedad a que habían venido las labores de minas, los frecuentes extravíos de piñas, etc." (BAN Minas t. 60, item 11), f. 1, statement by Roiz, los Lipes, June 18, 1703.

87. "Obra tan heroica . . ."—don Bartolomé González Poveda to the king, La Plata, August 20, 1682, [on verso] "Informa los méritos del maestre de campo Antonio López de Quiroga" (AGI Charcas 128).

88. Marquis of Montesclaros to the king, Lima, April 3, 1611 (AGI Lima 36), para. 3.

89. *Poder* of Antonio López to don Francisco de Robledo, don Juan Mazón, don Agustín Ponce de León, *agente de negocios*, and don Diego Suárez de Valdés, Potosí, March 6, 1669 (PCM EN 121, f. 214 14v.).

90. *Poder*, Potosí, March 7, 1669, of López de Quiroga and Rivera y

Quiroga to the *Contador* Juan Esteban de la Parra, of Lima, to seek any available positions (*oficios* and *cargos*) for them before the viceroy (PCM EN 121, ff. 215–16v.).

91. "No. 2" [and, on reverse] "El Capitán Antonio López de Quiroga, vecino de la Villa Imperial de Potosí [to] excelentísimo señor" (AGI Charcas 128). This document is not dated. References in it (ff. 4v.–5), however, to López's adits at Porco, Ocurí, and Aullagas, suggest that it was written in or after 1670; and the use of "Capitán" in the title limits the dating to before mid-1672, since by then López held the honorary military rank of *maestre de campo* (see discussion below in this chapter).

92. Memorandum "No. 1" to the king, marked on the reverse "Cámara, a 10 de junio de 1676" (AGI Charcas 128).

93. Richard Konetzke, "La formación de la nobleza en Indias," in *Lateinamerika. Entdeckung, Eroberung, Kolonisation. Gesammelte Aufsätze von . . .* (Cologne, 1983), pp. 360–61. This is an article originally published in *Estudios Americanos*, vol. 10 (Seville 1951), pp. 329–57.

94. López's report "No. 2" of the early seventies (see n. 91 above), f. 4v. The statement of his assets and services given in this paragraph is a combination of his own assertions in report "No. 2" and of those he made in 1674 in his letter "7" to the king dated Potosí, December 31, 1674 (AGI Charcas 128), and in the "Instrucción que el maestre de campo Antonio López de Quiroga, vecino, azoguero, y hacendado en la Villa de Potosí, envía al Reverendísimo Padre Comisario General de Indias Fray Antonio de Somoza, don Agustín Ponce, don Diego Suárez de Valdés, y don Juan de Ugarte, caballero del orden de Santiago, residentes en la villa de Madrid, para que gobernándose por ella sin faltar en cosa alguna, ajusten la pretensión que se expresa con las capitulaciones siguientes" (AGI Charcas 128). This document is undated, but the internal reference in it to December 1673 places it after then; and similarities of content with manuscripts from 1674 suggest strongly that it belongs to that year.

There is little doubt that some of the claims that López made about his silver production in these various representations to the crown were exaggerated. A total production of 40,000,000 pesos between when he started producing silver (say 1660) and 1674 implies an average of some 2,667,000 pesos a year, on which the royalty, at a fifth, would be about 533,000 pesos. But that is more than five times the annual royalty payment he claims to have made in the early seventies. Can the Amoladeras mines have yielded so extraordinarily highly in the sixties as to account for the discrepancy?

95. "El mero mixto imperio, jurisdicción alta y baja de horca y cuchillo. . . ." See "Instrucción . . ." of 1674, para 2. For the precise connotations of this legal definition, see Alfonso María Guilarte, *El Régimen señorial en el siglo XVI* (Madrid 1962), pp. 117–35.

96. "Instrucción . . ." of 1674, para. 4.

97. Konetzke, "La formación de la nobleza," p. 360.

98. García Carraffa, *Enciclopedia*, vol. 85, p. 114.

99. Letter "7" of December 31, 1674 (AGI Charcas 128). López deputized Somoza and three other agents in Madrid to represent him in his quest for a title in two documents: a *poder* issued in Potosí in 1674 (precise date unknown, since the end of the manuscript is missing) (PCM EN 126, f. 540–40v.), and the "Instrucción . . ." of 1674 (see n. 94 above). For the office of Franciscan Comisario General de Indias, see Schäfer, *El Consejo Real*, vol. 2, pp. 229–37. His job was to supervise Franciscan activities in the American colonies, and act as liaison between the hierarchy of the Order and the secular administration of the Indies.

100. (On reverse) "La Plata. Duplicado a su magestad. 1682. Don Bartolomé González Poveda . . . 20 de agosto, informa los méritos del maestre de campo Antonio López de Quiroga" (AGI Charcas 128).

101. Clarence H. Haring, *The Spanish Empire in America* (New York 1947), pp. 23–25.

102. In fact, this was also a further attack on the authority of the high court in La Plata, since the crown had delegated patronage in Charcas to the court. See *Recopilación* 1680, 1.6.26.

103. Haring, *The Spanish Empire*, p. 24.

104. References here to *encomenderos* and the institution of *encomienda* are, as any student of Spanish American history will instantly see, extremely fragmentary and possibly misleading. But to have attempted even a brief account of the working and the historical significance of *encomienda* would have unbalanced the text. Readers wishing to know more about the institution might start with Haring, *The Spanish Empire* . . . , ch. 3; and go for more detail to Silvio A. Zavala *La encomienda indiana* (2d. ed., Mexico City 1973), and Robert T. Himmerich, *The encomenderos of New Spain, 1521–1555* (Ph.D. dissertation, University of California, Los Angeles, 1984). One assumption that has been made in the text here should be noted: that López hoped to be able to appoint *encomenderos* of the sort that had existed in Spanish America in the early colonial decades—that is, men who actually administered Indi-

ans and collected tributes directly from them. After the mid-sixteenth century the trend was for *encomenderos* to become separated from the Indians nominally under their control, and to become mere recipients, via the royal Treasury, of part of the tribute payments collected from those Indians. Given the broad tenor of López's requests, however, and the fact that he was thinking of creating *encomiendas* in newly-incorporated "wild" territory, it seems likely that he had in mind *encomenderos* on the old pattern, who would truly govern a group of native people and collect tributes, in cash or in kind, directly from them. This, however, is surmise.

105. Konetzke, "La formación de la nobleza," pp. 360–61.

106. Ibid., p. 360.

107. Miguel Herrero García, *Ideas de los españoles del siglo XVII* (Madrid 1966), ch. 7.

108. With one exception: in 1680, a *fiscal*, or attorney, of the Council of the Indies, summarizing a report and petition by López, wrote "A lo que se reduce la pretensión de don Antonio es a pedir merced de Conde de la provincia de Pilaya y Pazpaya . . ." (AGI Charcas 128, note by the *fiscal* of August 31, 1680, attached to a letter from don Benito de Rivera y Quiroga to don Juan de Austria, Potosí, October 28, 1679). Whether this was a mere slip of the pen, or whether the attorney automatically thought that anyone of López's wealth and accomplishments must inevitably be a *don*, is an open question.

Some light may be shed on this generally puzzling matter by the fact that López's father, Alvaro de Quiroga y Rivera, and one of his uncles, Pedro López de Quiroga, also were without the *don*; although another uncle, Juan de Quiroga, who was the one to serve as guard captain to the Viceroy of Sicily, did possess the title. These distinctions reinforce the suggestion that gradations of standing existed *within* the Quiroga family. See Paredes's preface to la Gándara's *Las palmas y triunfos . . . ,* f. 2.

109. See *poder* of Potosí, May 11, 1690, issued by López to don Luis Francisco de Madrigal and others (PCM EN 138, f. 90–90v.). The letter of 1679, dated September 28, is in AGI Charcas 128.

110. Konetzke, "La formación de la nobleza," p. 361. See also Doris M. Ladd, *The Mexican nobility at Independence, 1780–1826* (Austin 1976), pp. 15–17.

111. Bravo's success also shows that wealth and economic benefit to the state were not enough, for though he was the leading miner of Zacatecas in his day, his operations and output were trivial in compari-

son to López's. See P. J. Bakewell, *Silver mining and society in colonial Mexico. Zacatecas, 1546–1700* (Cambridge: Cambridge University Press, 1971), p. 119.

112. For the equivalence, see Juan Marchena Fernández, *Oficiales y soldados en el ejército de América* (Seville 1983), p. 69. No document has appeared conveying López's appointment, but it is in 1672 that general manuscripts in Potosí begin to refer to him with his new rank. The first to do so is Viceroy Lemos's "Provisión del gobierno en que se adjudica al maestre de campo Antonio López de Quiroga el socavón y minas que se adjudicaron para su magestad en el asiento de Laicacota a don Antonio de Andrade," Lima, February 6, 1672 (PCM CR 418, f. 142v.).

113. Arzáns, *Historia*, vol. 2, p. 267.

114. Roberto Levillier, *El Paitití, el Dorado, y las Amazonas* (Buenos Aires 1976), pp. 93, 282.

115. Ibid., p. 267.

116. For this, see John Hemming, *The conquest of the Incas* (New York, 1970).

117. Levillier, *El Paitití*, p. 101.

118. Ibid., p. 270.

119. Toledo to Philip II, 1572 (no exact date given), ibid., p. 270.

120. Report of Juan Recio de León, ibid., p. 272.

121. Ibid., pp. 275, 277.

122. Don Benito de Rivera y Quiroga and Antonio López de Quiroga to the crown, Potosí, March 16, 1670 (AGI Charcas 23). The account given in this paragraph of the beginnings of López's venture of the Gran Paitití draws on this letter, another by the same writers to the Council of the Indies, of Potosí, March 14, 1670 (AGI Charcas 23), and a section in the middle of López's letter "7" to the crown, of Potosí, December 31, 1674 (AGI Charcas 128).

123. PCM EN 121, ff. 211–12. Fernández's grant had been issued by the Viceroy Count of Alba de Aliste.

124. López's letter "7" of December 31, 1674.

125. Ibid.

126. Ibid.

127. López and Rivera to "excelentísimo senor," Potosí, March 14, 1670. This was apparently the first that the Council of the Indies had heard of their project. On June 5, 1671, it instructed Viceroy Lemos to report on the significance of the enterprise, and, if appropriate, give it all favor and support (AGI Charcas 23).

128. Arzáns, *Historia*, vol. 2, p. 273. Antonio López was not yet, of

course, a *maestre de campo* at the time when this show of arms took place. But Arzáns, writing, as he did, after López's death, generally attributes the title to López throughout his lifetime.

129. Letter "7" of December 31, 1674. López bought 1,000 mules from the province of Buenos Aires for the purposes of the expedition. Most were to be sold at a profit to cover costs, and some were sent to Rivera y Quiroga for use as freight animals. See López's power of attorney to Father Juan de Soria, prior of the Augustinian house in the town of Mizque, north of Potosí, dated Potosí, December 30, 1670 (PCM EN 123, ff. 644–45).

130. López's letter "7" of December 31, 1674.

131. Ibid. The river is not named. It is likely to have been one of the headwaters of the Mamoré.

132. Ibid.

133. Ibid.: "Buena gente para romper montañas . . . ."

134. Ibid.

135. "Da cuenta a Vuestra Magestad don Benito de Rivera y Quiroga, gobernador y capitán general de la conquista del Gran Paitití, del estado en que se halla en ella." Report marked "9" on first page, Potosí, October 28, 1679 (AGI Charcas 128).

136. Ibid., and a shorter letter, marked "11," of very similar phrasing, Rivera y Quiroga to the king, Potosí, November 28, 1679 (AGI Charcas 128).

137. Rivera's report "9" of October 28, 1679.

138. Royal *cédula* of Madrid, November 25, 1680 (BAN *cédulas reales* No. 528).

139. "Informa la Real Audiencia de La Plata sobre el estado de la conquista del Gran Paitití del cargo del maestre de campo Antonio López de Quiroga y don Benito de Rivera y Quiroga," La Plata, August 21, 1682 (AGI Charcas 25).

140. "1696. Don Benito de Rivera y Quiroga, gobernador del Gran Paitití, sobre lo que el maestre de campo Antonio López de Quiroga . . . se obligó a dar para la prosecución de la conquista de dicha provincia" (BAN Minas t. 19, item 9 [Minas catalog No. 1,134]), ff. 2–3, *Petición* of Rivera y Quiroga, contained in a *Real Provisión* of La Plata, September 17, 1696.

141. Ibid., f. 6.

142. Arzáns, *Historia*, vol. 2, p. 274. This passage was written in or before 1708 (ibid., vol. 1, p. xvi), and to judge from Arzáns's use of the present tense, Rivera y Quiroga was still alive then.

143. Marqués del Saltillo, *Linajes de Potosí* (Madrid 1949), p. 52. The concession was by a *"real despacho"* of November 19, 1686.

144. In two large testimonials of his services drawn up in 1689 and 1690, López makes very little of the Paitití episode. In fact the document of 1690 does not mention it at all (Antonio de Valenzuela to "muy poderoso señor," Potosí, April 1, 1690, in AGI Charcas 128). And the other stresses the evangelizing aim, and is silent on both the heroic qualities and the economic potential of the explorations that feature so largely in López's initial writings about the Paitití, twenty years earlier. López now just reiterates how much the venture has cost him, notes that communication over the mountains has been established, and reports that Indian villages discovered are now evangelized by Dominicans from Cochabamba, Franciscans from Apolobamba, and Jesuits from Santa Cruz. See the *Información* by and on Antonio López, final date Potosí, January 21, 1690 (AGI Charcas 128), f. 2v.

145. BAN Minas t. 16, item 1 (Minas catalog No. 1,003), ff. 81v.–82, 86.

146. "1675. El maestre de campo Antonio López de Quiroga con Bartolomé de Villalobos sobre el derecho a la veta nombrada San Francisco de Asís . . . en el asiento de Berenguela, provincia de Pacajes" (BAN Minas t. 67, item 5 [Minas catalog No. 999]), f. 20.

147. Lohmann Villena, *El Conde de Lemos*, p. 459.

148. Henry Charles Lea, *A History of the Inquisition in Spain* (4 vols., New York 1922), vol. 2, pp. 243–44.

149. For Velasco's titles, see his *recibo de dote*, Potosí, October 21, 1676 (PCM EN 128, ff. 245–50v.).

150. Ibid., f. 248v.

151. For the complete dowry list, ibid., passim.

152. PCM EN 128, f. 245.

153. PCM EN 129, f. 308v.

154. Arzáns, *Historia*, vol. 2, p. 395.

155. Ibid., p. 398.

156. Don Bartolomé González Poveda, president of the high court of La Plata, to king, La Plata, July 26, 1681, "Informe general de los sujetos beneméritos eclesiásticos y seculares" (AGI Charcas 24).

157. *Poder* of López to Velasco, Potosí, September 11, 1677 (PCM EN 129, ff. 148–51).

158. An inventory of his goods at his death is given in PCM EN 145, f. 259. For his knighthood, Arzáns, *Historia*, vol. 2, p. 461.

159. The date of the marriage has not surfaced. It probably took place in the early eighties, since in 1710 the couple's eldest son, Captain don Francisco de Gambarte y Quiroga, appears in litigation as the guardian and administrator of his younger siblings. To exercise such a rôle, he was presumably past the age of majority—that is, twenty-five years old or more. He would have been born, then, in 1685 or earlier. See "1709–1711. El capitán Blas Míguez, dueño del socavón de San Blas, cerro de Potosí, sobre el quinto de los metales que doña María y doña Lorenza López de Quiroga . . . han sacado de sus labores, desde el año de 1690 por dicho socavón" (BAN Minas t. 30, No. 7), ff. 29–32.

160. *Residente* in a power of attorney that he granted in Potosí, on May 28, 1676 to Captain don Andrés Vázquez de Velasco and don Pablo Vázquez de Velasco, of Lima, to represent him for all purposes before the Viceroy Conde de Castellar. For this, see an unfoliated sewn file of manuscripts marked on the back "Plata. 1 de diciembre de 1677. El señor Presidente don Bartolomé González de Poveda. Da cuenta a Vuestra Excelencia que el pleito del comiso queda revistado en lo principal . . ." in AGI Charcas 129. The author is indebted to Dr. Zacarías Moutoukias for pointing out the references to Gambarte in this *legajo*. For the description of Gambarte as *vecino* and *capitán*, see his purchase of a black slave woman, Potosí, September 26, 1676 (PCM EN 128, ff. 203–4v.).

161. ". . . donde tiene su mayor consumo."—*Memorial* of Gambarte addressed to "excelentísimo senor," undated, but possibly 1676, in the file "Plata. 1 de diciembre de 1677. El señor Presidente don Bartolomé González de Poveda . . ." (see previous note). Another visitor to the Río de la Plata, the French traveller and merchant Acarete du Biscay, who passed through Buenos Aires a few years before Gambarte, in 1658–59 and again in the early sixties, confirms the heavy consumption of *mate* in the Andean mines.

"Without this herb (with which they prepare a refreshing drink with water and sugar, which should be taken warm) the inhabitants of Peru, both natives and others, and especially those who work in the mines, could not survive; for the earth is full of mineral veins, and the vapors that rise from these would be suffocating, and nothing could restore [the miners] except this infusion, which brings them to life again and gives them back their former strength."

Great quantities of *mate* tea were taken by miners when they left the mines to eat or rest. See Acarette, *Relación de un viaje al Río de la*

Plata y de allí por tierra al Perú. Con observaciones sobre los habitantes,
sean indios o españoles, las ciudades, el comercio, la fertilidad, y las
riquezas de esta parte de América, tr. Francisco Fernández Wallace (Buenos Aires: Alfer & Ways, 1943), pp. 33–4, 77.

162. The suit, with accompanying reports from local authorities, occupies the whole of AGI Charcas 129.

Smuggling through Buenos Aires was big business by this time.
The Spanish home authorities were well aware of this, though not
able to control the trade to any degree; and so any merchant dealing
in the Río de la Plata was the object of much suspicion. On his arrival in 1658, Acarete du Biscay had found no fewer than twenty-two
Dutch boats lying off Buenos Aires, loading hides, apparently with
the connivance of the governor. See *Relación,* pp. 45, 95. Business
could be most profitable. Acarete calculated his own and his partners'
profits on their dealings at Buenos Aires (selling mostly European
cloths, and iron and steel goods, and taking on mainly hides and vicuña
wool for sale in Europe), at a net 250 percent. (Ibid., p. 93). He also
tells how he smuggled silver aboard his ship—evidently not a difficult
feat. (Ibid., p. 87.)

163. "Cámara [of the Consejo de Indias] en 10 de septiembre 1681.
Don Joseph de Gambarte en virtud de poder de don Miguel de Gambarte,
su hermano," at the end of a long, water-damaged, file entitled "Año de
1676. Cuaderno 21. Copia de los autos y edictos, cargos de los ausentes,
y escritos del fiscal en esta averiguación y pesquisa que hace por particular comisión de su Magestad el maestre de campo don Andrés de Robles
. . ." (AGI Charcas 60).

164. Arzáns, *Historia,* vol. 2, p. 295, n. 4.

165. PCM EN 134, f. 119.

166. Shipping contract of Potosí, August 31, 1678 (PCM EN 130, ff.
542–24v.).

167. Captain Miguel de Vergara—to whom Gambarte sent 81 bars of
silver and 19,500 pesos in coin from Potosí, May 7, 1681 (PCM EN 133,
ff. 273–74v.). A bar at this date usually weighed around 160 marks (80
pounds), and would be worth, at that weight, about 1,280 pesos. So the
total shipment sent by Gambarte amounted to 120,000 pesos or so.

168. See Gambarte's power-of-attorney to López and Juan de Eche-
verría, of Potosí, June 6, 1687, to collect debts, take mines, buy land,
and litigate (PCM EN 136B, f. 559–59v.).

169. Arzáns, *Historia,* vol. 2, p. 436.

CHAPTER 4

1. For the election of January 1, 1676, see BAN, Cabildo de Potosí, Libro de Acuerdos [CPLA] vol. 30, f. 142; for that of 1678, CPLA vol. 31, f. 113v.

2. Lohmann Villena, *El Conde de Lemos*, p. 459.

3. The purchase, of three mines in the Santa Elena vein, was from doña Luisa and doña Petronila Vázquez de Ayala y Figueroa, and was transacted on July 16, 1677. See PCM EN 129, ff. 138–39v.

4. PCM EN 136, ff. 292–93v.

5. For the purchase, Potosí, July 8, 1681, see PCM EN 133, ff. 192–93v.

6. *Recopilación* 1680, 5, 2, 47.

7. *Compañía* formed in Potosí, October 24, 1686 to open an adit "Jesus Nazareno" at Ocurí (PCM EN 136, ff. 292–93v.)

8. *Información* on Antonio López de Quiroga of Potosí, January 21, 1690 (AGI Charcas 128), ff. 7, 15v.

9. *Fianza*, Potosí, January 2, 1677 (BAN CPLA vol. 17, f. 18v.). The possibility of a Galician connection rests on a similarity of names. Araujo is an uncommon surname; and so, although there is no clue to don Joseph's origins, the appearance of another Araujo in Potosí in the mid-1670s who definitely was Galician is suggestive. This was the *maestre* don Juan Araujo, a clergyman who arrived from Spain about 1676, and who got to know Antonio López "on account of being from the same homeland." López made him chaplain of one of his refineries on the *Ribera* of Potosí. See the *Información* on López of Potosí, April 1, 1690 (AGI Charcas 128), no foliation, Araujo's testimony of La Plata, January 5, 1690.

10. The power-of-attorney (*poder para muchos efectos*) was issued in Potosí on November 11, 1679. See PCM EN 131, ff. 462–65v.

11. Arrazola died in 1679, and the office, the *alcaldía provincial* of the Santa Hermandad, was bought by don Pedro Ponce de León in October of that year. See BAN CPLA vol. 31, ff. 248v.–50.

12. For don Francisco de Bóveda y Saravia, see the *parecer* on him of the high court of La Plata, November 26, 1674 (BAN, Audiencia de Charcas, Libros de acuerdos, vol. 13, f. 232–32v.). It was to don Francisco in Potosí that the *corregidor* of Lipes, don Cristóbal de Quiroga y Osorio, wrote on August 29, 1677, announcing the successful junction of Antonio López's adit at San Antonio with mines on the Veta Rica there (the letter is attached to another from the Viceroy Count of Castellar to the king, Lima, March 15, 1678, in AGI Charcas 23).

13. For the chaplaincy (*capellanía*), of August 30, 1677, see PCM EN 129, ff. 308–9v. The capital on which it was founded was a large dwelling house, with attached yard (*cancha*), belonging to López in the "barrio del matadero de vacas" of Potosí. He valued the house at 6,000 pesos. At the normal interest rate of 5 percent for such transactions ("*de veinte mil el millar*"), the *capellanía* yielded 300 pesos.

14. "Aquellas dos familias que ya vimos a quinientos que andaban tan hermanadas que parecían una misma. . . ." (quoted in González López, *La Galicia de los Austrias*, vol. 1, p. 373).

15. On July 28, 1668, he issued a guarantee on behalf of don Martín Choque, an Indian official of the community of Atunquillacas responsible for delivering *mita* Indians to Potosí, to the effect that Choque would return to Potosí after making a journey with Quiroga. See PCM EN 121, f. 51–51v.

16. See the discussion of the agreement in Ch. 3 above. Lemos's appointment of Quiroga is noted in Lohmann Villena, *El Conde de Lemos*, p. 458.

17. Two years later, however, on October 4, 1674, one *maestre de campo* don Cristóbal Quiroga y Sotomayor received the title of *corregidor* of los Lipes from the Viceroy Count of Castellar (BAN CPLA vol. 30, ff. 30v.–36). Since this is the only reference so far revealed by research to anyone with this name, it is possible that Quiroga y Osorio and Quiroga y Sotomayor were the same man, and that the appointment of *corregidor* was simply being renewed. Sotomayor is an ancient Galician family name (see García Carraffa, *Enciclopedia*, vol. 85, p. 197 ff.). In either case, Antonio López seems likely to have been able to count on useful backing from a kinsman in los Lipes during the mid seventies.

18. *Poder* issued by Antonio López, Potosí, June 3, 1678 (PCM EN 130, ff. 190–92v.).

19. See BAN Minas t. 58, item 5 (Minas catalog No. 1,042), f. 118v.

20. BAN Minas catalog No. 1,057, f. 16v.

21. On April 30, 1682, Antonio López paid 500 pesos into the Potosí treasury towards a total of 2,000 owed by Quiroga for a four-year lease of these mines and a refinery. The lease ran from March 1, 1679. See PCM CR 360, f. 84.

22. PCM EN 138, ff. 68–70v.

23. BAN Minas t. 17, item 5 (Minas catalog No. 999), passim.

24. BAN CPLA vol. 31, ff. 82v.–93v. The Count of Castellar issued the title in Lima, on August 15, 1677.

25. The appointment was made by the *corregidor* of Potosí, don Pedro

Luis Enríquez, who was certainly no enemy of Antonio López. See his letter to the king, Potosí, September 11, 1684 (AGI Charcas 60). See the title pages of the account books numbered PCM CR 469, 472, 477, and 478 for evidence of Quiroga's tenure of office from 1683 to 1686.

26. See the "Relación de las misiones y conversión de los infieles del Gran Paitití," made before don Diego de Quiroga at Cochabamba, January 10, 1688 (AGI Charcas 128).

27. "MDCLXX años. Data de lo pagado de Hacienda Real por salarios ordinarios . . .," in PCM CR 426, ff. 110–120v., indicates that he was in office in 1669. The salary of the *protector* was 600 pesos a year. He is mentioned as the incumbent of the office in 1673 in an *Obligación* acknowledged in Potosí on May 10 of that year by don Pablo Cayo to Antonio López de Quiroga (PCM EN 125, folio not recorded).

28. See, in general, *Recopilación*, book 6, title 6.

29. Title issued in Lima, June 2, 1679 (BAN CPLA vol. 31, ff. 292–96).

30. See *Prologue*.

31. On October 25, 1675, Antonio López despatched to Lima 84,000 pesos: 10,000 of his own, and 74,000 belonging to Liñán y Cisneros, who at that time was still in La Plata (PCM EN 127, f. 542–42v.). And on September 30, 1679, López sent off to Liñán in Lima 24,241 pesos in unassigned church income (*tercias vacantes*) that he had had collected in the see of Charcas (PCM EN 131, f. 388–88v.). By this time Liñán was archbishop in Lima and also serving as interim viceroy.

32. "Información" by the *alcaldes veedores del Cerro* on *socavones* and *barrenos* begun in the Rich Hill during the administration of don Pedro Luis Enríquez, *corregidor* of Potosí, Potosí, November 22, 1686 (AGI Charcas 61). The inspection showed that since Enríquez had come into office, in 1678, nine new adits had been started, including López's at the Amoladeras, and extension of six older galleries had been undertaken, one of them being the adit at Polo that López had acquired. Perhaps, therefore, the example of López's success with adits in and outside Potosí had stimulated imitation.

33. Testimony of don Diego de Figueroa y Paz, Potosí, November 22, 1689, f. 5 in López's *Información* of Potosí, January 21, 1690 (AGI Charcas 128).

34. Potosí, March 17, 1689, "Testimonio de la vista de ojos que a pedimento del maestre de campo Antonio López de Quiroga se ejecutó por el alcalde mayor de minas del Cerro Rico de Potosí de los cinco socavones que en dicho cerro ha dado el dicho maestre de campo" (AGI Charcas 128).

35. Ibid.
36. Ibid.
37. Ibid.
38. Ibid.
39. Don Diego de Figueroa y Paz, *alguacil mayor* of the Tribunal of the Santa Cruzada in Potosí, testimony of November 22, 1689, f. 7v. in López's *Información* of Potosí, January 21, 1690.
40. *Informe* of Enríquez (now Conde de Canillas de Torneros), Potosí, December 30, 1689. Ibid., f. 73v.
41. I.e., taxes on silver production. Ibid., f. 3v.
42. Araujo's testimony, La Plata, January 5, 1690, in the report on López prepared by Antonio de Valenzuela, final date, Potosí, April 1, 1690 (AGI Charcas 128); unfoliated.
43. Ibid., Pardo de Figueroa's testimony, La Plata, January 7, 1690.
44. Tomás García Ramón, testimony of December 5, 1689, f. 42v. in the *Información* of January 21, 1690.
45. *Fray* Buenaventura de Madrid, *guardián* of the Franciscan house in La Plata, testimony at La Plata, January 7, 1690, for Valenzuela's report.
46. López to the king, Potosí, September 28, 1679 (AGI Charcas 128).
47. Don Diego de Figueroa y Paz, f. 5v. of the *Información* of January 21, 1690. These were nine *cabezas de ingenio*, or individual trituration devices. Since a refinery might have two such machines in it (though never more), López probably had a minimum of five refineries (*ingenios*) in Potosí.

In Ocurí he still had two *ingenios* running in 1690 (ibid., f. 7). And in San Antonio he operated at least one, which had been partly or totally rebuilt in the late eighties. He issued a power-of-attorney in Potosí, May 17, 1690, to don Joseph de Ujúe y Gambarte, nephew of Captain Miguel de Gambarte, to administer a new water-driven refinery in San Antonio (PCM EN 138, ff. 316–18).

48. A *certificación* (Potosí, September 30, 1695) by Diego Pérez de Lazcano, accountant of an inspection of the Treasury office in Potosí carried out by the *corregidor*, don Pedro Luis Enríquez, 1695, notes that López "poseyó" three *ingenios* in the town formerly belonging to Juan Guillén, don Antonio Osores de Ulloa, and Juan Rodríguez de Villapalma. The use of the past tense may indicate that López no longer owned these refineries. See AGI Charcas 25.

49. That formerly belonging to Francisco Carrión de la Serna, for which López still owed money to the Treasury. See the statement of the Potosí Treasury officials, May 4, 1696 (PCM CR 360, ff. 92v.–93).

50. PCM CR 360, ff. 92, 94.

51. Don Diego de Figueroa y Paz stated in November 1689 that López was then extracting 3 marks, or 24 ounces, of silver from each *cajón* of 50 *quintales* of ore refined. This amount of silver was worth, he said, 20 pesos (evidently after tax); whereas López's refining costs for a *cajón* of ore were 50 pesos. (For this statement, f. 10v. of the *Información* on López of January 10, 1690). How representative these figures were, or how carefully chosen to add lustre to López's efforts, is impossible to say. Obviously, though, if the ore yield quoted was anywhere close to reality, López was certainly working with far poorer ores than he had extracted from the Amoladeras mines in the sixties. See above, Ch. 2. For other references to poor ore quality at this time, see Antonio de Valenzuela's report on López of April 1, 1690—Valenzuela's own opening statement, and the evidence of Dr. don Pedro de Velasco y Velázquez, La Plata, January 3, 1690.

52. Culpina was an *estancia* where López bred mares (*yeguas*)—more probably for mules than for horses. To judge from references in "1692–1694. José Cipriano de la Cruz, indio residente en la hacienda de Culpina . . . contradiciendo el yanaconaje a que pretende reducirlo el maestre de campo Antonio López de Quiroga. . . ." (BAN Minas t. 19, item 6 [Minas catalog No. 1,LL8]), f. 2, López had been the owner of the land for the previous twenty-four years or more.

53. See, for a reference to Ingahuasi, the *Obligación* granted in Potosí, September 2, 1670, by the *Alférez Real* Francisco Moreira Calderón to López for the delivery of cattle to Ingahuasi in 1672 (PCM EN 123, f. 383–83v.).

54. "Venta de la mitad y dos partes de la estancia de Chaquilla," by doña María Núñez Cortés and others to Antonio López de Quiroga, Potosí, August 8, 1682 (PCM EN 132, ff. 357–60v.).

55. Sale of the *estancia y tierras* of Chaquilla by Antonio López to Andrés de Salazar, Potosí, February 20, 1688 (PCM EN 137, ff. 50–53v.). The total sale price of 26,890 pesos 5 reales also included another piece of land, called Cotani, 1 by 3 leagues in area, and an abandoned silver refinery.

56. Sale by the *albaceas* of Espinosa Patiño to Antonio López of Carapari, described as *haciendas de viñas y tierras*, and another nearby *hacienda de viña, estancia, y tierras*, Potosí, May 20, 1688 (PCM EN 137, ff. 418–21v.).

57. For these landholdings, see BAN Minas t. 19, item 8 [Minas catalog No. 1,140]; "Concierto y convenio entre el capitán don Miguel de

Gambarte y su muger y doña Lorenza de Quiroga y entre don Santiago de Ortega y su muger," Potosí, March 9, 1700 (PCM EN 148, ff. 64–65v.); and "Poder y nombramiento de administrador y mayordomo: doña Lorenza de Quiroga a don Miguel de Vargas y Ocampo," Potosí, January 26, 1699 (PCM EN 147, ff. 14–16v.).

58. *Poder* of Antonio López to don Joseph de Asáldegui, Potosí, August 29, 1696 (PCM EN 144, f. 403–03v.)

59. *Poder* of Antonio López to Diez de Quiroga and others, Potosí, November 7, 1692 (PCM EN 141, ff. 109–110v.).

60. The principal administrator seems to have been Antonio López Martínez. For the *poder* to him, Boada, and Diez de Quiroga, of Potosí, August 29, 1695, see PCM EN 143, ff. 179–80v.

61. See n. 52 above.

62. These figures are calculated from accounts presented in a suit entitled "1690–1691. Doña Juana de Tovalina, viuda del sargento mayor Francisco de Lemos y Uzcátegui, que fue receptor de alcabalas en Potosí, sobre siete mil y tantos pesos que en razón de ese derecho quedó debiendo el maestre de campo Antonio López de Quiroga. . . ." (BAN Minas t. 19, item 3 [Minas catalog No. 1,101]), especially ff. 4–18.

63. See "1697–1698. El maestre de campo Antonio López de Quiroga sobre las alcabalas que Martín López de Recalde y Mateo Rodríguez de Espinosa . . . pretenden cobrarle por los frutos de sus haciendas" (BAN Minas t. 19, item 12 [Minas catalog No. 1,143]), ff. 10–11v.

64. Ibid., ff. 16–20.

65. "1690–1691. Doña Juana de Tovalina . . . ," ff. 4, 21.

66. For example, Antonio de Valenzuela in his report on López, Potosí, April 1, 1690. López's mining efforts are supported by the estates he holds in Pilaya y Pazpaya "whose fruits he has converted into the working of mines, since those of the mines themselves do not correspond to the many expenses he has had in working mines, in supplying refineries, in the adits he has made, and in providing supplies to other refiners."

67. BAN Minas t. 70, item 3 [Minas catalog No. 1,029], f. 19v.

68. Pedro de Cieza de León, *La crónica del Perú* (BAE, vol. 26, Madrid 1947), ch. 96.

69. *Historia*, vol. 2, p. 268.

70. Ibid.

71. See López's *Obligación* to Vázquez de Velasco, Potosí, July 1, 1678 (PCM EN 130, f. 303–3v.). Arzáns (ibid.) gives the weight of a basket (*cesto*) of coca as one *arroba*, or roughly 25 pounds.

72. Felipe Guaman Poma de Ayala, *El primer nueva corónica y buen gobierno* (ed. John V. Murra and Rolena Adorno, Mexico City 1980), vol. 1, p. 251.

73. *Historia*, vol. 2, p. 269.

74. Agreement between López and Captain don Francisco de Lemos de Aller y Usátegui, former *receptor de alcabalas*, Potosí, June 6, 1686 (PCM EN 136, f. 138). For López's earlier remittances to Vázquez de Castro, see PCM EN 131, f. 174–74v.; PCM EN 134, f. 54–54v.; PCM EN 135, ff. 35–35v., 173–73v.; PCM EN 136, f. 146–46v. This record is not necessarily complete because of gaps in notarial series.

75. 24,241 pesos despatched from Potosí on September 30, 1679 (PCM EN 131, f. 388–88v.).

76. 3,160 pesos despatched from Potosí on January 13, 1677 (PCM EN 129, f. 512–12v.). The monastery was of the Order of Saint Francis of Paola. One *Fray* Tomás de Cárdenas belonging to it had died in the Mataca valley.

77. Despatch of 16,178 pesos on December 1, 1678, to Captain Diego Pérez Lobo in Lima, whose *ropa de la tierra* López had handled and sold. See PCM EN 130, f. 129–29v.

78. For this remittance, Potosí, March 17, 1673, see PCM EN 125, f. 157–57v.

79. *Obligación* of the *alférez real* Francisco Moreira Calderón, resident of the province of Buenos Aires, to López, to supply this number of *vacas y novillos* at 20 *reales* a head, Potosí, September 2, 1670 (PCM EN 123, f. 383–83v.).

80. In 1675 López granted a power-of-attorney to agents in Buenos Aires to recover the advance payment that he had made for the cattle, which he had still not received. See his *poder* to Captain Juan Domínguez and other residents of Buenos Aires, Potosí, May 21, 1675 (PCM EN 127, ff. 578–79v.).

81. *Deudo y obligación* by López and Miguel de Gambarte, Potosí, August 31, 1696, to the cathedral of Buenos Aires and its agent in Potosí, Captain Juan Ruiz de la Fuente. López had received 4,979 animals for Ingahuasi, and Gambarte 3,376 head for his *hacienda* of Oploca, in the province of Chichas. (This estate had formerly belonged to don Alvaro Espinosa Patiño de Velasco, and had evidently been bought at some point by Gambarte.) See PCM EN 144, ff. 384–85.

82. See, for example, the *poder* issued by López and Gambarte to Captain Antonio de la Tijera, dated Potosí, September 12, 1692, empowering him to spend up to 20,000 pesos in Buenos Aires on any available

goods imported from Spain or locally produced. Iron was the only item specifically mentioned. (PCM EN 141, f. 105–05v.).

83. *Licenciado* Bartolomé González Poveda to the king, La Plata, January 30, 1678, "No. 2" (AGI Charcas 129).

84. Ibid. The prices therefore refer to 1678.

85. Ibid.

86. Azañón to the king, La Plata, October 18, 1692 (AGI Charcas 62), f. 1–1v.

87. "Certificación de cómo el maestre de campo Antonio López de Quiroga es síndico de la religión de Nuestro Padre San Francisco," Potosí, January 26, 1689 (AGI Charcas 128).

88. Ibid.

89. Spain pressed strongly in the seventeenth century for proclamation as Catholic dogma of the notion that Mary was free of original sin on account of her own immaculate conception. Though the dogma was not in fact established by Rome until 1854, the Church instituted the feast of the Immaculate Conception nearly two hundred years earlier, in 1664. See Edward J. Sullivan and Nina A. Mallory, *Painting in Spain, 1650–1700, from North American collections* (Princeton: The Art Museum, Princeton University, 1982), p. 91.

90. See the *Concierto de retablo*, Potosí, June 23, 1675, between Miguel de Ortega, *maestro del oficio de ensamblador* on the one hand, and López, as the *prioste* of the brotherhood, and Captain Antonio Arias, as its *mayordomo*, on the other (PCM EN 127, ff. 628–29v.).

91. This was presumably the altar piece contracted for in 1675 (see previous note). If so, it was an imposing object, nearly 28 feet (10 *varas*) in height, and costing 2,100 pesos.

92. See his *Certificación* of January 1689 (n. 87 above).

93. *Historia*, vol. 2, p. 398. Arzáns is mistaken in the month. The entry for López's death in the book of burials for the period still to be found in the cathedral archive in Potosí records that he was buried on January 24, 1699. See Archivo de la Catedral (Potosí), *Libro de Entierros* No. 2, f. 196v.

CONCLUSION

1. *Fray* Buenaventura de Madrid, *guardián* of the Franciscan monastery of La Plata, La Plata, January 7, 1690, in Antonio de Valenzuela's report on López, Potosí, April 1, 1690 (AGI Charcas 128).

2. See Chapter 2 above.

3. 500 *quintales* of a total of 2,256 *quintales* and 39 pounds distributed. The next largest amount, 100 *quintales*, went to López's partner at San Antonio del Nuevo Mundo, don Alvaro Espinosa Patiño. Excluding López, thirty-five refiners bought mercury from the Treasury in 1684. Their average purchase was of almost exactly 50 *quintales*—a tenth of López's amount. See "Razón de lo que se debe a su magestad en su Real Caja de la Villa Imperial de Potosí por todos efectos . . . No. 1. Azogue que se debe del que se repartió para la armada de este año de 1675" (AGI Contaduría 1,815, No. 1).

4. AGI Contaduría 1,815, No. 5, ff. 260–79v.—"Data" of "Azogues en especie."

5. Ibid., ff. 260–66v. The *veinticuatro* Alonso Martínez de Uceda bought 110 quintales, and don Pedro Urrutigoiti Echauz took 109.

6. Captain Gaspar de Mariaca, Potosí, July 15, 1690, "Certificación de las cantidades que el maestre de campo Antonio López de Quiroga ha pagado de quintos desde principios de 1673 hasta fin del de 1690 [actually, June 30, 1690]—item appended to Valenzuela's report on López (see n. 1 above). Mariaca gives the weight of the bars as 2,186,298,553 *maravedís*. Converted at the standard contemporary rate of 2,380 to the mark, this is equivalent to the number of marks given here in the text.

7. Bakewell, "Registered silver production," Table 1, p. 96.

8. *Historia*, vol. 2, p. 395.

9. Bakewell, "Registered silver production," Table 1, pp. 95–96.

10. This estimate uses the assumption that López continued to produce a seventh or eighth of Potosí's total output between 1691 and his death. There is no evidence either to contradict or support this projection of the share he had contributed over the previous thirty years.

11. *Historia*, vol. 2, p. 394.

12. An undated *Memorial* to the king on mining prospects in Charcas, by don Juan de Lizarazu, president of the Audiencia of La Plata, 1633–42 (AGI Charcas 20).

13. For a fuller discussion of the objections, see Chapter 2.

14. See Chapter 2.

15. See the discussion of López's work in San Antonio del Nuevo Mundo in Chapter 3.

16. López to the king, Potosí, September 28, 1679 (AGI Charcas 128).

17. *Licenciado* don Bartolomé González Poveda to the king, La Plata, August 20, 1682 (AGI Charcas 128).

18. Bakewell, "Registered silver production," graph 1, facing p. 86.

19. "Testimonio de la vista de ojos . . .," Potosí, March 17, 1689 (AGI Charcas 128).

20. Bakewell, *Miners of the red mountain*, pp. 128–30.

21. *Historia*, vol. 2, p. 398.

22. López was also without a doubt the single largest employer in Potosí, putting into the Rich Hill, according to two witnesses, as many workers as all the other miners combined. (Testimony of *fray* Buenaventura de Madrid and don Ignacio Pardo de Figueroa, both in La Plata, January 7, 1690, for Antonio de Valenzuela's report on López, Potosí, April 1, 1690 [AGI Charcas 128]).

23. At Porco he said he had spent 300,000 pesos on his adit by 1681; at Ocurí, 200,000 pesos on two adits by 1689; at San Antonio del Nuevo Mundo, almost 218,000 pesos between 1672 and 1677 (the amount given by his *minero*, Alonso Ruiz, and accepted by López). (See Chapter 3 above for these figures). To these outlays must be added his expenditure on adits at Laicacota, and, on a far larger scale, on the many he drove into the Hill of Potosí. Arzáns estimated that López had put two million pesos into his adits (*Historia*, vol. 2, p. 397).

24. Dr. don Pedro Vázquez de Velasco, treasurer of the cathedral in La Plata and vicar general of that archdiocese, in testimony given at La Plata, January 3, 1690, for Antonio de Valenzuela's report on López, Potosí, April 1, 1690 (AGI Charcas 128).

25. "Las piadosas entrañas . . .": Don Ignacio Pardo de Figueroa Montenegro, testimony at La Plata, January 7, 1690, in Valenzuela's report on López, Potosí, April 1, 1690 (AGI Charcas 128). Pardo, a *gallego*, had been one of López's *mineros* in Potosí.

26. Joseph Favio Gutiérrez, testimony at Potosí, December 23, 1689, in the *Información* on López of Potosí, January 21, 1690 (AGI Charcas 128), f. 69v.

27. This was another *maestre de campo*, Vicente de Zaldívar, a prominent figure not only in Zacatecas itself, but across the north of New Spain. See P. J. Bakewell, *Silver mining and society in colonial Mexico. Zacatecas, 1546–1700* (Cambridge: Cambridge University Press, 1971), p. 46.

28. La Borda, an immensely enterprising silver producer at Tlalpujahua, Taxco, and Zacatecas in the middle decades of the eighteenth century, was one of several Mexican miners in that century whose careers resemble that of Antonio López. He was Spanish, born in Aragon on January 2, 1699 (three weeks almost to the day before López's death), and crossed to New Spain as a youth, about 1716. See Modesto Bargalló,

La minería y la metalurgía en la América española durante la época colonial (Mexico City: Fondo de Cultura Económica, 1955), pp. 286–87.

29. *Historia*, vol. 2, p. 396. Arzáns picks up the same theme on p. 397. "As for his charity toward the poor, some say that he fell short (and many agree with this), although others say that he did give to many beggars, 200 pesos to one, 500 or 1,000 to others. . . ."

30. *Historia*, vol. 2, p. 398.

31. *Historia*, vol. 2, p. 394.

32. *Historia*, vol. 2, p. 398.

33. Students of entrepreneurship will recognize these qualities of alertness to gainful opportunity, willingness to accept uncertainty and risk, openness to technical change, and capacity to see and realize new combinations of productive and marketing processes, including rationalization of them, as traits often ascribed to entrepreneurs, or considered to constitute entrepreneurship. Authorities consulted on the subject include R. H. Campbell and R. G. Wilson (eds.), *Entrepreneurship in Britain, 1750–1939* (London 1975); Arthur Seldon (ed.), *Prime mover of progress. The entrepreneur in capitalism and socialism* (London: The Institute of Economic Affairs, 1980); Mark Casson, *The entrepreneur. An economic theory* (Oxford: Martin Robertson, 1982); Israel M. Kirzner, "The primacy of entrepreneurial discovery" (in Seldon, ed., *Prime mover of progress*); and, of course, fundamental to all consideration of entrepreneurs, Joseph A. Schumpeter, *The theory of economic development. An inquiry into profits, capital, credit, interest, and the business cycle* (Oxford: Oxford University Press, 1961).

34. G. Micheal Riley, *Fernando Cortés and the Marquesado in Morelos, 1522–1547. A case study in the socioeconomic development of sixteenth-century Mexico* (Albuquerque: University of New Mexico Press, 1973), especially Chapter 6.

35. James Lockhart, "Encomienda and hacienda: The evolution of the great estate in the Spanish Indies," *Hispanic American Historical Review*, 49:3 (August 1969), pp. 411–429; and the same author's *Spanish Peru, 1532–1560. A colonial society* (Madison: University of Wisconsin Press, 1968), especially Chapter 2.

36. François Chevalier, *La formation des grands domaines au Mexique. Terre et société aux XVIe–XVIIe siècles* (Paris: Institut d'Ethnologie, 1952), p. 162.

37. Chevalier, *La formation*, pp. 190–93.

38. Franklin Pease G. Y., "Relaciones entre los grupos étnicos de la Sierra sur y la costa: Continuidades y cambios," in Luis Millones and

Hiroyasu Tomoeda (eds.), *El hombre y su ambiente en los Andes centrales* (Osaka: Senri Ethnological Studies 10, 1982), p. 109. See also the earlier work referred to here by Pease: Rómulo Cuneo Vidal, "El Cacicazgo de Tacna," originally published in 1919, and reprinted in the same author's *Obras completas* (Lima 1977), vol. 1, pp. 316–67.

39. See John V. Murra, "Aymara lords and their European agents at Potosí," *Nova Americana*, 1 (Turin 1978), pp. 231–43.

40. Kirzner, "The primacy of entrepreneurial discovery," p. 17.

41. The clearest statement of the *peruleros'* methods and successes, and at the same time one of the strongest cases made to date for the growth of economic autonomy in seventeenth-century Spanish America, is Margarita María Suárez Espinosa, *Las estrategias de un mercader: Juan de la Cueva, 1608–1635* (Lima: Pontificia Universidad Católica del Perú, Facultad de Letras y Ciencias Humanas, Memoria para obtener el grado de Bachiller en Humanidades con mención en Historia, 1985). The first part of this thesis analyzes the *peruleros'* methods. The second part traces the career of one of the most successful of their number, Juan de la Cueva, who was not only a merchant but a prominent banker in Lima.

42. Bakewell, *Silver mining and society*, pp. 115–16.

43. Bakewell, *Silver mining and society*, pp. 118–19, 135.

44. Henry Kamen, *Spain in the later seventeenth century, 1665–1700* (London 1983), p. 266. See in general the discussion from p. 260 onward.

45. For these restraints—for instance, Alfonso X's pronouncement in the thirteenth century that a noble's position was compromised by any "personal engagement in commerce or any low manual occupation to earn money"—see William J. Callahan, *Honor, commerce and industry in eighteenth-century Spain* (Boston 1972), p. 3 ff.

46. Kamen, *Spain in the later seventeenth century*, p. 263.

47. William J. Callahan, "Don Juan de Goyeneche: Industrialist of eighteenth-century Spain," *Business History Review*, 43:2 (Summer 1969), pp. 154–55.

48. D. A. Brading, *Merchants and miners in Bourbon Mexico, 1763–1810* (Cambridge: Cambridge University Press, 1971), passim.

49. Richard L. Garner, "Silver production and entrepreneurial structure in 18th century Mexico," *Jahrbuch für Geschichte von Staat, Wirtschaft und Gesellschaft Lateinamerikas*, 17 (1980), pp. 157–85.

50. Richard J. Salvucci, "Entrepreneurial culture and the textile manufactories in eighteenth-century Mexico," *Anuario de Estudios Americanos*, vol. 39 (Sevilla, 1982), pp. 397–419. John E. Kicza, *Colonial*

entrepreneurs. Families and business in Bourbon Mexico City (Albuquerque: University of New Mexico Press, 1983).

51. Arzáns, *Historia*, vol. 2, pp. 423–44.

52. On this point, for the final decades of the eighteenth century, and for the mines of Peru, see John R. Fisher, *Silver mines and silver miners in colonial Peru, 1776–1824* (Liverpool: University of Liverpool, Centre for Latin American Studies, 1977), pp. 33–34. For Potosí, see Rose Marie Buechler, *The mining society of Potosí, 1776–1810* (Syracuse: Department of Geography, Syracuse University, 1981), especially pp. 275–90. Buechler estimates that the three leading *azogueros* of Potosí at the end of the eighteenth century, Luis de Orueta, Antonio Zabaleta, and Pedro Antonio Azcárate, together produced some 7,000,000 pesos of silver between 1780 and 1805. On average over those twenty-five years each therefore produced some 10,700 marks annually. Antonio López's average annual production over the thirty-seven years from 1661 to 1698 was, to take a low estimate, some 44,000 marks annually— that is, considerably more than the combined output of Orueta, Zabaleta, and Azcárate. (See the calculations of López's lifetime silver production at the beginning of this *Conclusion*.)

53. One possible reason is the advantage of cheap labor that employers in New Spain may have enjoyed with the rapid growth of all segments of the Mexican population in the eighteenth century. Demographic increase in the Andes, on the other hand, was slower, the Indian population not having reached its post-Conquest minimum there until late in the seventeenth century, several decades after its Mexican counterpart. Mexico also gives the impression of far greater general prosperity than the central Andes after 1700, which suggests that capital for mining may have been easier to raise there than in the Andes. For the reluctance of merchants to invest in mining in Peru, see Fisher, *Silver mines*, chapter 6, and especially p. 98 ff.

SUMMARY OF MANUSCRIPT SOURCES

INFORMATION FOR THIS BOOK HAS BEEN TAKEN LARGELY FROM SPANISH and Bolivian archives. Three repositories proved equally valuable in their different ways: the archive of the Casa Nacional de Moneda in Potosí (PCM), for notarial and treasury records; the Archivo Nacional de Bolivia in Sucre (BAN), for mining litigation; and the Archivo General de Indias in Seville (AGI), for official correspondence and other sundry matters. Particularly valuable in Seville is *legajo* Charcas 128, which, though nominally devoted to the Treasury's pursuit of Antonio López for his refusal to pay sales taxes on agricultural products brought by him to Potosí, in fact ranges far wider than that over his activities.

The following is a list of the archival volumes or bundles (*legajos*) cited in the notes. Many others were searched, but proved to contain nothing of interest for the topics at hand.

AGI:
Section—Audiencia de Charcas, *legajos* 17, 20, 21, 23, 24, 25, 28, 32, 35, 36, 46, 52, 60, 61, 62, 123, 128, 129, 134.
—Audiencia de Lima, *legajos* 35, 36, 37, 45, 52, 53, 54.
—Contaduría, *legajo* 1815.
—Contratación, *legajo* 5,419.
—Escribanía de Cámara, *legajo* 865C.

BAN:
Section—Audiencia de Charcas, libros de acuerdos, fol. 13.
—Cabildo de Potosí, libros de acuerdos, vols. 17, 30, 31.
—Minas, vols. 12, 16, 17, 19, 30, 56, 58, 60, 67, 70, 97, 135, 136, 142, 144.

MANUSCRIPT SOURCES

PCM:

Section—Archivo de la Casa Real de Moneda, *caja* 6, *legajo* 1.

    —Cajas Reales, vols. 20, 207, 264, 316, 321, 339, 360, 361, 384, 385, 393, 399, 402, 408, 411, 415, 418, 419, 423, 426, 445, 469, 472, 477, 478, 503.

    —Escrituras notariales, vols. 53, 88, 112, 113, 116, 118, 119A, 121, 123, 124, 125, 126, 127, 128, 129, 130, 131, 132, 133, 134, 135, 136, 136B, 137, 138, 141, 143, 144, 145, 147, 148.

In addition to the above, these manuscripts are cited in notes:

Real Academia de la Historia (Spain), Colección Muñoz, ms A/70.

Archivo General de la Nación (Argentina), Sala 13, Cuerpo 23, 102 Cuaderno 7.

BIBLIOGRAPHY OF WORKS CITED

Acarette [du Biscay]. *Relación de un viaje al Río de la Plata y de allí por tierra al Perú. Con observaciones sobre los habitantes, sean indios o españoles, las ciudades, el comercio, la fertilidad, y las riquezas de esta parte de América.* Trans. by Francisco Fernández Wallace. Buenos Aires: Alfer & Ways, 1943.

Acosta, José de. *Historia natural y moral de las Indias, en que se tratan* [sic] *de las cosas notables del cielo, elementos, metales, plantas, y animales dellas, y los ritos y ceremonias, leyes y gobierno de los indios (1590).* Ed. by Edmundo O'Gorman. 2d. ed., Mexico City, 1962.

Alonso Barba, Alvaro. *Arte de los metales, en que se enseña el verdadero beneficio de los de oro y plata por azogue.* Madrid, 1630. Repr., Potosí: Colección de la Cultura Boliviana, 1967.

Arzáns de Orsúa y Vela, Bartolomé. *Historia de la Villa Imperial de Potosí.* 3 vols. Ed. by Lewis Hanke and Gunnar Mendoza L. Providence: Brown University Press, 1965.

Bakewell, Peter. *Miners of the red mountain. Indian labor in Potosí, 1545–1650.* Albuquerque: University of New Mexico Press, 1984.

———. "Mining in colonial Spanish America." In *The Cambridge History of Latin America.* Vol. 2. *Colonial Latin America.* Ed. by Leslie Bethell. Cambridge: Cambridge University Press, 1984.

———. "Registered silver production in the Potosí district, 1550–1735." *Jahrbuch für Geschichte von Staat, Wirtschaft und Gesellschaft Lateinamerikas,* 12 (1975), 67–103.

———. *Silver mining and society in colonial Mexico. Zacatecas, 1546–1700.* Cambridge: Cambridge University Press, 1971.

———. "Technological change in Potosí: The silver boom of the 1570s."

Jahrbuch für Geschichte von Staat, Wirtschaft und Gesellschaft Lateinamerikas, 14 (1977), 60–77.

Bargalló, Modesto. *La amalgamación de los minerales de plata en Hispanoamérica colonial.* Mexico City: Compañía Fundidora de Fierro y Acero de Monterrey, 1969.

———. *La minería y la metalurgía en la América española durante la época colonial.* Mexico City: Fondo de Cultura Económica, 1955.

Barnadas, Josep M. "Una polémica colonial: Potosí, 1579–1584." *Jahrbuch für Geschichte von Staat, Wirtschaft und Gesellschaft Lateinamerikas*, 10 (1973), 16–70.

Brading, D. A. *Merchants and miners in Bourbon Mexico, 1763–1810.* Cambridge: Cambridge University Press, 1971.

Buechler, Rose Marie. *The mining society of Potosí. 1776–1810.* Syracuse: Syracuse University, Department of Geography, 1981.

Callahan, William J. "Don Juan de Goyeneche: Industrialist of eighteenth-century Spain." *Business History Review*, 43:2 (Summer 1969), 152–70.

———. *Honor, commerce, and industry in eighteenth-century Spain.* Boston, 1972.

Campbell, R. H., and R. G. Wilson (eds.). *Entrepreneurship in Britain, 1750–1939.* London, 1975.

Capoche, Luis. *Relación general de la Villa Imperial de Potosí.* Ed. by Lewis Hanke. Biblioteca de Autores Españoles, vol. 122. Madrid: Atlas, 1959.

Casson, Mark. *The entrepreneur. An economic theory.* Oxford: Martin Robertson, 1982.

Chevalier, François. *La formation des grands domaines au Mexique. Terre et société aux XVIe–XVIIe siècles.* Paris: Institut d'Ethnologie, 1952.

Cieza de León, Pedro de. *La crónica del Perú.* Biblioteca de Autores Españoles, vol. 26. Madrid: Atlas, 1947.

Cole, Jeffrey A. *The Potosí mita, 1573–1700. Compulsory Indian labor in the Andes.* Stanford: Stanford University Press, 1985.

Contreras, Jaime. *El Santo Oficio de la Inquisición de Galicia (poder, sociedad y cultura).* Madrid, 1982.

Crespo Rodas, Alberto. *La Guerra entre vicuñas y vascongados, Potosí 1622–1625.* 2d. ed. La Paz: Colección Popular, 1969.

Dodge, Meredith E. *Silver mining and social conflict in seventeenth-century Peru: Laicacota, 1665–1667.* Albuquerque: University of New Mexico, Ph.D. dissertation, 1984.

Enciclopedia universal ilustrada europeo-americana. 80 vols. Madrid: Espasa Calpe, 1908–1933.

Fisher, John R. *Silver mines and silver miners in colonial Peru, 1776–1824.* Liverpool: University of Liverpool, Centre for Latin American Studies, 1977.

Frisancho, A. R. "Human growth and development among high-altitude populations." In P. T. Baker (ed.), *The biology of high-altitude peoples.* Cambridge: Cambridge University Press, 1978.

García Carraffa, Alberto and Arturo. *Enciclopedia heráldica y genealógica hispano-americana.* 88 vols. Madrid, 1952–63.

Garner, Richard L. "Silver production and entrepreneurial structure in 18th century Mexico." *Jahrbuch für Geschichte von Staat, Wirtschaft und Gesellschaft Lateinamerikas,* 17 (1980), 157–85.

González López, Emilio. *La Galicia de los Austrias.* Tomo 1. *1506–1598.* La Coruña: Fundación "Pedro Barrie de la Maza, Conde de Fenosa," 1980.

Guaman Poma de Ayala. Felipe. *El primer nueva corónica y buen gobierno.* Ed. by John V. Murra and Rolena Adorno. Mexico City: Siglo XXI, 1980.

Guilarte, Alfonso María. *El régimen señorial en el siglo XVI.* Madrid, 1962.

Hakluyt, Richard. *Voyages and discoveries. The principal navigations, voyages, traffiques and discoveries of the English nation.* Ed. by Jack Beeching. Harmondsworth: Penguin Books, 1972.

Haring, Clarence H. *The Spanish Empire in America.* New York: Oxford University Press, 1947.

Hemming, John. *The conquest of the Incas.* New York, 1970.

Herrero García, Miguel. *Ideas de los españoles del siglo XVII.* Madrid: Gredos, 1966.

Himmerich, Robert T. *The encomenderos of New Spain, 1521–1555.* Los Angeles: University of California, Ph.D. dissertation, 1984.

Hooson, William. *The miner's dictionary, explaining not only the terms used by miners, but also containing the theory and practice of that most useful art of mining, more especially of lead mines.* Wrexham, 1747. Repr. by The Institution of Mining and Metallurgy, London 1979.

Howe, Walter. *The Minining Guild of New Spain and its Tribunal General, 1770–1821.* Cambridge, Mass.: Harvard University Press, 1949.

Jiménez de la Espada, Marcos. *Relaciones geográficas de Indias—Perú,* vol 1. Biblioteca de Autores Españoles, vol. 183. Madrid: Atlas, 1965.

Kamen, Henry. *Spain in the later seventeenth century. 1665–1700.* London: Longman, 1983.

Kicza, John E. *Colonial entrepreneurs. Families and business in Bourbon Mexico City.* Albuquerque: University of New Mexico Press, 1983.

Konetzke, Richard. "La formación de la nobleza en Indias." In *Lateinamerika. Entdeckung, Eroberung, Kolonisation. Gesammelte Aufsätze von* . . . . Cologne, 1983.

Ladd, Doris M. *The Mexican nobility at Independence,* 1780–1826. Austin: Institute of Latin American Studies, 1976.

La Gándara, Fray Felipe de. *Nobiliario, armas, y triunfos de Galicia, hechos heroicos de sus hijos, y elogios de su nobleza, y de la mayor de España, y Europa, compuesto por el Padre Maestro* . . ., *de la Orden de San Agustín, Coronista General de los Reinos de León y Galicia.* Madrid, 1677.

Lea, Charles Henry. *A History of the Inquisition in Spain.* 4 vols. New York, 1922.

Levillier, Roberto. *El Paitití, el Dorado, y las Amazonas.* Buenos Aires: Emecé, 1976.

———. *Gobernantes del Perú. Cartas y papeles, siglo XVI.* Vol. 8. *Ordenanzas del Virrey Toledo.* Madrid, 1925.

Lockhart, James. "Encomienda and hacienda: The evolution of the great estate in the Spanish Empire." *Hispanic American Historical Review.* 49:3 (August, 1969), 411–429.

———. *Spanish Peru. A colonial society.* Madison: University of Wisconsin Press, 1968.

Lohmann Villena, Guillermo. *El Conde de Lemos, virrey del Perú.* Madrid: Escuela de Estudios Hispanoamericanos de Sevilla, 1946.

Maffei, Eugenio, and Ramón Rúa Figueroa. *Apuntes para una biblioteca española de libros, folletos y artículos, impresos y manuscritos, relativos al conocimiento y explotación de las riquezas minerales y a las ciencias auxiliares.* 2 vols. Madrid, 1871. Repr. by VI Congreso Internacional de la Minería, León, 1970.

Marchena Fernández, Juan. *Oficiales y soldados en el ejército de América.* Sevilla: Escuela de Estudios Hispanoamericanos, 1983.

Mendoza L., Gunnar. *Guerra civil entre vascongados y otras naciones de Potosí.* Potosí: Cuadernos de la Colección de la Cultura Boliviana, 1954.

BIBLIOGRAPHY

Monge, Carlos. *Acclimatization in the Andes. Historical confirmations of "Climatic Aggression" in the development of Andean man.* Baltimore: The Johns Hopkins Press, 1948.

Morse, Richard M. "The urban development of colonial Spanish America." In *The Cambridge History of Latin America.* Vol. 2. *Colonial Latin America.* Ed. by Leslie Bethell. Cambridge: Cambridge University Press, 1984.

Murra, John V. "Aymara lords and their European agents at Potosí." *Nova Americana*, 1 (Turin, 1978), 231–43.

Padden, Robert C., ed. *Tales of Potosí. Bartolomé Arzáns de Orsúa y Vela.* Trans. by Frances M. López-Morillas. Providence: Brown University Press, 1975.

Pease G. Y., Franklin. "Relaciones entre los grupos étnicos de la Sierra sur y la costa: Continuidades y cambios." In Luis Millones and Hiroyasu Tomoeda (eds.). *El hombre y su ambiente en los Andes centrales.* Osaka: Senri Ethnological Studies 10, 1982.

Recopilación de leyes de los Reynos de las Indias. Mandadas imprimir, y publicar por la Magestad Católica del Rey Don Carlos II. 4 vols. Madrid, 1681. Facsimile ed., Madrid: Ediciones Cultura Hispánica, 1973.

Riley, G. Micheal. *Fernando Cortés and the Marquesado in Morelos. 1522–1547. A case study in the socioeconomic development of sixteenth-century Mexico.* Albuquerque: University of New Mexico Press, 1973.

Rudolph, William E. "The lakes of Potosí." *The Geographical Review*, 26:4 (New York, October 1936), 529–54.

Ruiz Almansa, Javier. *La población de Galicia, 1500–1945.* Madrid: Consejo Superior de Investigaciones Científicas, 1948.

Saltillo, Marqués de. *Linajes de Potosí.* Madrid, 1949.

Salvucci, Richard J. "Entrepreneurial culture and the textile manufactories in eighteenth-century Mexico." *Anuario de Estudios Americanos*, 39 (Sevilla, 1982), 397–419.

Schäfer, Ernst. *El Consejo Real y Supremo de las Indias. Su historia, historia, organización y labor administrativa hasta la terminación de la Casa de Austria.* Tomo II. *La labor del Consejo de Indias en la administración colonial.* Sevilla: Escuela de Estudios Hispanoamericanos, 1947.

Schumpeter, Joseph A. *The theory of economic development. An inquiry into profits, capital, credit, interest, and the business cycle.* Oxford: Oxford University Press, 1961.

BIBLIOGRAPHY

Seldon, Arthur (ed.). *Prime mover of progress. The entrepreneur in capitalism and socialism*. London: The Institute of Economic Affairs, 1980.

Suárez Espinosa, Margarita María. *Las estrategias de un mercader: Juan de la Cueva, 1608–1635*. Lima: Pontificia Universidad Católica del Perú, Facultad de Letras y Ciencias Humanas, Memoria para obtener el grado de Bachiller en Humanidades con mención en Historia, 1985.

Sullivan, Edward J., and Nina A. Mallory. *Painting in Spain, 1650–1700, from North American collections*. Princeton: The Art Museum, Princeton University, 1982.

Zavala, Silvio A. *La encomienda indiana*. 2d ed., Mexico City: Porrúa, 1973.

INDEX

Acosta, José de, 91
adits, *Fig. 3*, 75; at Aullagas,
77–78; at Berenguela, 83; at
Chayanta, 77; in las
Amoladeras, 61–62; at Ocurí,
77, 78; at Laicacota, 73; at
Porco, 74–76; in Rich Hill of
Potosí in 1680s, 135–139; at
San Antonio del Nuevo
Mundo, 77, 86–92, 96; at Titiri,
80–81; at Tomahavi, 83; driven
by blasting, 76; *socavón de
Benino*, 91; utility of, 62,
160–161
agricultural products, from
Antonio López's lands, 142–144
altitude adaptation, 188 n. 34
Amoladeras, las, 55; mined by
Antonio López, 60–63, 65,
135–136, 139, 159–160; laborers
in, 136; ores, 63
Araujo, don Juan, 138, 221 n. 9
Araujo y Gayoso, *sargento mayor*
don Joseph de, *corregidor* of los
Lipes, 129
Arrazola, Francisco de, 130
Arzáns de Orsúa y Vela,
Bartolomé: on coca, 145–146
Aullagas: Antonio López active
at, 77–78

aviador, 49
avío. See credit
Aymaya, 207 n. 38
Azañón y Velasco, *general* don
Gregorio, 3–5, 134, 149
Azurza, Ignacio de, 85, 93, 139

Baeza, 147
Benino, Nicolás del, 91, 175
Berenguela, 83, 131, 132
Berrío site, 65, 136
Biloma, 120
blasting: benefits of, 76–77, 163;
cartridges, 91; in Europe, 205
n. 17; at Huancavelica, 205 n.
17; at Ocurí, 79; used first in
the Potosí district at Porco,
76–77; at San Antonio del
Nuevo Mundo, 91–92; at Titiri,
for mines, 81; source of
powder, 210 n. 62
Boada y Quiroga, don Benito de,
142
Borda, José de la, 168, 230 n. 28
Bóveda, Lorenzo de, and family,
13–15, 51, 109, 157, 187 n. 27
Bóveda, doña María de, 61
Bóveda y Saravia, doña Agustina
de, 130